Intelligent User Interface

This book aims to mainstream the UI/UX design process by explaining the latest AI/ML models in a comprehensible way and highlighting case studies on developing intelligent user interfaces for extended reality (XR) systems, human–robot interaction, cockpit design and trajectory prediction. The book also discusses the latest standards and guidelines relevant to UI/UX design, layout and equipment list for setting up a lab on intelligent interaction design involving robots, drones and XR systems.

- Covers a wide array of topics ranging from human factors, computer vision, AR/VR systems, large language models (LLMs) and usability evaluation techniques, including statistical hypothesis techniques
- Discusses the latest AI systems such as vision transformers, LLM-based human–robot interfaces and virtual reality-based spacecraft simulation systems
- Provides a list of free downloadable software on the covered topics
- Contains graphical illustrations and a list of quick facts for easy review and recall of basic concepts in each chapter
- Gives new project ideas on intelligent user interfaces that can be explored by students and early career researchers

The intended audience of this book is engineering and design students and faculty members, user interface designers and product managers who would like to be aware of the latest AI/ML without diving into too many theoretical details, and use it for their project or product development.

Intelligent User Interface

Usable Artificial Intelligence and Artificial Intelligence for Usability

Pradipta Biswas

CRC Press
Taylor & Francis Group
Boca Raton London New York

CRC Press is an imprint of the
Taylor & Francis Group, an **informa** business

Designed cover image: Pradipta Biswas

First edition published 2026
by CRC Press
2385 NW Executive Center Drive, Suite 320, Boca Raton FL 33431

and by CRC Press
4 Park Square, Milton Park, Abingdon, Oxon, OX14 4RN

CRC Press is an imprint of Taylor & Francis Group, LLC

ISBN: 9781041087335 (hbk)
ISBN: 9781041087342 (pbk)
ISBN: 9781003646716 (ebk)

DOI: 10.1201/9781003646716

Typeset in Sabon
by Deanta Global Publishing Services, Chennai, India

Contents

Preface

Why should you read this book? There is a popular statement that 'Experts do not know everything, but they know where to look'. In today's era, the statement can be rewritten as 'Experts do not know everything, but they know the appropriate keywords for Google search or the best prompt for ChatGPT'. On a more serious note, a developer, designer, manager or any stakeholder for a new development should be able to pinpoint the pain point of novelty – what is the new invention, why it did not exist so far, why it works or may not, how a technology can be made more efficient and so on.

Artificial intelligence and machine learning (AI/ML) and user interface and user experience (UI/UX) design are well-known disciplines with a plethora of offline and online resources. This book aims to mainstream UI/UX design by presenting relevant theories of AI with case studies of intelligent user interfaces from a wide range of applications. The book covers topics ranging from basic human factors, AI/ML, computer vision, natural language processing (NLP), large language models (LLMs), human–robot interaction, cockpit design, trajectory prediction and standardization. Each chapter is carefully drafted to introduce readers to the historical aspects, current trends and relevance to interface or interaction design with a list of five bullet points as quick facts for ease in remembering. None of the topics are discussed in too many details, and all basic concepts and terms are explained to ensure accessibility for readers with different backgrounds. Each chapter contains links to multiple lecture notes and video lectures to dive into the details of individual topics. Chapter 9 on human–robot interaction also discusses sample codes to develop LLM-enabled human–robot interfaces for rovers and manipulators.

In the context of electronic product or service design, success depends on user acceptability. In the recent trends of using or incorporating AI into the product life cycle, the success of AI will depend upon the usability of the product or service. Similarly, the usability of emerging technology such as novel modalities of interaction or augmented and virtual reality (AR/VR) and LLM systems can be improved by the appropriate use of AI technology. Overall, this book aims to help designers, engineers and managers to identify appropriate technology for novel product or service development and efficiently channel resources through an overall understanding of technology related to intelligent interface design.

About the Author

Pradipta Biswas is an associate professor in the Department of Design and Manufacturing, associate faculty in the Robert Bosch Centre for Cyber Physical Systems and Curriculum Committee member of HAL-IISc Skill Development Centre at the Indian Institute of Science. His experiment on human–computer interaction is one of the seven experiments selected for the Axiom Mission 4 launched in 2025. He was elected as vice chairman of ITU Study Group 9 (2022–2024) and also as co-chair of the IRG AVA (2020–2024) and Focus Group on Smart Cable Television of the International Telecommunication Union, the telecom branch of the United Nations. He is a convenor of the BIS (Bureau of Indian Standards) Panel on Metaverse.

His research focuses on user modelling and multimodal human–machine interaction for aviation and automotive environments and for assistive technology. He set up and led the Interaction Design (I3D) Lab. His research won Microsoft's AI 4 Accessibility Grant Award, Facebook's Responsible AR/VR Award and research grants from Collins Aerospace (on AR-AI integration), Siemens (on holoportation), British Telecom (on AR/VR systems), Faurecia Groupe Services Ltd. (on automotive UI), Wipro Ltd (on autonomous vehicle), Defence Research and Development Organization (DRDO), Indian Space Research Organization (ISRO), Departments of Science & Technology (DST) and BioTechnology (DBT). He was featured in *New Scientist*, NDTV, All India Radio and various other media outlets.

Biswas was a senior research associate with the Engineering Department, research fellow at Wolfson College and research associate at Trinity Hall of the University of Cambridge. He completed a PhD in computer science at the Rainbow Group of the University of Cambridge Computer Laboratory and Trinity College in 2010 and was awarded a Gates-Cambridge Scholarship in 2006. He is a member of the UKRI International Development Peer Review College, and was a professional member of the Society of Flight Test Engineers and British Computer Society and an associate fellow at the UK Higher Education Academy and Royal Society of Medicine.

Acknowledgements

The book was conceived during the COVID pandemic when we lost in-person classes and had to rely on online resources for teaching. I started developing and maintaining a webpage indexing my lecture notes and video recordings of lectures. After the lockdown period, a new question emerged on the necessity of a physical book. The internet, social media and online repositories are overwhelmed with AI resources, and for a student or practitioner, it often became difficult to separate the good and correct material from jargon. Thanks to my students and their feedback, I felt the need for a concise textbook introducing the latest concepts of AI to design students, industrial practitioners and managers without diving into too much detail on individual topics, rather enabling one to identify the pain points in current practices or products and select the correct topic for detailed investigation.

I would like to thank my present and past students and the following people for actively contributing to the drafting effort:

- Mr Yashaswi Sinha and Mr Mukund Mitra for reviewing Chapters 5 and 6, helping in drafting use cases for Chapter 11 and preparing a figure explaining recent NLP models
- Dr Somnath Arjun for his help in drafting Chapter 7 on visualization
- Dr Abhishek Mukhopadhyay for drafting the table differentiating CNNs from vision transformers in Chapter 5 and supplying figures for the model validation part in Chapter 2
- Dr LRD Murthy for his help in making Figure 4.1 and drafting the aviation case study in Chapter 10
- Mr Himanshu Vishwakarma and Ms Arushi Khokhar for their help in drafting use cases for Chapter 9
- Mr Naveen Talwar for his help in drafting the use case for Chapter 11
- Mr Maharudra Kharsade for writing captions for figures in Chapters 2 and 3
- Mrs JayaLakshmi helped on preparing the figures for production

Besides them, I took snippets from the doctoral and master's theses of Prof Gowdham Prabhakar, Dr Subin Raj, Dr Vinay Krishna Sharma and Mr Ajit Krishnan. Finally, I would like to thank my parents and family for their support in completing the draft.

Chapter 1

Introduction

1.1 BACKGROUND

What is intelligence? One of many possible definitions can be 'goal-directed adaptive behaviour'. While the 2024 Nobel Prizes recognized the work in the field of artificial intelligence (AI), the term itself is relatively new, first appearing during the Dartmouth Workshop in 1956. However, one of the primary aims of AI was and still is to make human-like machines or perhaps a virtual human, thanks to Hollywood movies, maybe a look-alike of the best actor of the day. This concept of a virtual human took root in the pursuit of developing automated machines in the historic era. While Chapter 3 will look back at the history of automation, this pursuit of a virtual human or human-like system brings AI closer to human–machine interaction. In the mid-1980s, when computers were gradually becoming a household commodity from a lab sophistication, the book *The Psychology of Human Computer Interaction* initiated a new discipline in human–computer interaction (Card et al. 1983). One of the earliest cognitive models, called the Model Human Processor (Card et al. 1983), and its more ambitious cousin, the SOAR cognitive architecture (Newell 1990), borrow their propositional logic-based knowledge representation from the General Problem Solver, sort of a holy grail of symbolic AI. Taking one step back, it was during the Second World War when theoretical analysis of man–machine interaction became important for the rapidly evolving military hardware, and soon a need was felt for a 'user model' or 'virtual human' to automatically test hardware against a range of human abilities. With the advancement of AI and computing hardware, intelligent user interfaces (IUIs) gradually became commonplace through new applications, interaction techniques, data interpretation and so on.

Now in the present era of Alphafold, GenAI and semi-autonomous warfare, what does it take to make an intelligent user interface? Should it be designed like traditional UI/UX storyboarding? Should it be trained like a classical supervised learning agent? Will it replace humans or co-operate or compete? Before diving into these very relevant questions, let us take a basic and innocent example of the challenges ahead. Let's consider an example of

DOI: 10.1201/9781003646716-1

a small child. Even before he turns a year old, he learns to recognize familiar faces and toys, can pick up and place objects of different shapes and sizes, can move either by crawling or walking towards a destination while avoiding obstacles and so on. Now what does it take for a group of engineers to make a robotic system that can pick and place objects of arbitrary shape and size, navigate while avoiding obstacles and so on? Of course, all these activities are already implemented, perhaps in separate systems, but the overall resource requirement is rather significant than trivial. Alan Newell, the first author of the book *The Psychology of Human Computer Interaction*, presented a paper at the 1972 Carnegie Symposium titled 'You can't play 20 questions with Nature and win', pointing out the fragmented research on understanding human cognition (Newell 1973) and the replication of them in machines. While traditional AI and computer vision took their own course on bio-inspired design, the field of intelligent user interface bridges the gap between AI and end user applications to make things intelligent and intuitive. From next word prediction in a virtual keyboard to humanoid robots, the intelligent user interface is both essential and a decisive factor for user acceptance of new interactive technology.

1.2 BRIEF DESCRIPTION

There are plenty of books on both AI/ML and UI/UX design, but the majority of books on intelligent user interfaces are edited volumes lacking the coherent structure of a textbook. Presenting the diverse range of topics related to UI/UX design and artificial intelligence in a single volume is ambitious; however, the present book will take readers through a rather horizontal journey of developing intelligent user interfaces with pointers to vertical dives for individual topics of AI and interaction design. The book will first help readers understand their end users in terms of describing the basics of cognitive psychology and user modelling. Chapter 2 will cover basic theories of visual search, visual pathways, colour perception and auditory perception. Next, different approaches to user modelling and interface personalization will be discussed. From user modelling and personalization, the book will move to present a few basic theories of AI and machine learning. The third chapter will present historical aspects of automation, robotics and computation, and in that context, present different types of knowledge representations used in artificial intelligence and different types of machine learning, albeit at a subjective level. With the knowledge of basic human factors and AI/ML, the book will take readers into novel interactive devices, including gaze-controlled interfaces, AR/VR displays, haptics and spatial audio. Chapter 5 will discuss recent computer vision models in terms of convolutional neural networks and vision transformers, followed by a brief discussion and case study on synthetic data generation. Chapter 6 will present basic theories of content analysis, hyperlink analysis and the structure of chatbots or recommender systems. In this context, introductory details on large language models (LLMs) and their application for IUIs will be discussed. The next chapter will look at different data visualization techniques, starting from basic visual primitives to complex visualization techniques through topological data analysis. Chapter 8 will describe usability evaluation techniques, controlled experiment design and statistical hypothesis testing. The next chapter will discuss human–robot interaction in different contexts, including CoBots for industrial applications and social robotics systems. The chapter will present two contemporary case studies on extended reality (XR) and LLM-based human–robot Interaction. Chapter 10 will describe representative case studies of intelligent user interfaces for cockpit design, which are often more challenging than consumer electronics systems. Case studies will cover automotive, aviation and spacecraft-related human–machine interfaces. In particular, an interactive heads-up display for cars, an eye gaze-controlled head-mounted display system for military aviation, a modular VR simulator for spacecraft and a robotic manipulator for children with severe speech and motor impairments. The chapter also introduces the important topic of cognitive load estimation for pilots and drivers. Chapter 11 will present another important type of intelligent interaction in the form of target prediction or intent recognition. It will present three case studies involving hand movement prediction, walking trajectory prediction and human–robot handover requests. The penultimate chapter will move from technology development to long-term sustainability through standardization. The chapter will discuss relevant standards from ISO, ITU, NHTSA and US DoD. Appendix 1 will present the layout and equipment list of a tentative lab design for undertaking research on intelligent user interfaces and sample questions. A series of lecture notes and video lectures is available at https://cambum.net/PB/lecture-notes.php. A list of software can be downloaded from the links given in Appendix 2 and sample questions can be found at Appendix 3.

In summary, the book will introduce topics of human factors and basic AI and in that context present novel interaction techniques and usability evaluation methods. Subsequent chapters will cover basic theories of information retrieval, visualization and human–robot interaction and applications for IUI. The ninth chapter will briefly describe applications ranging from assistive to aerospace technology. Chapter 12 will take users to the world of international standards, followed by a summary and concluding remarks.

1.3 OBJECTIVES

AI is now integrated into most interactive devices and applications, yet the topic of intelligent user interface is rarely taught as a separate course. Traditional courses on AI/ML or UI/UX design often miss either of the topics and do not cover case studies on IUI design and evaluation. This book will offer a condensed summary of relevant topics in AI/ML and UI/UX design and evaluation with case studies ranging from assistive technology to spacecraft HMI. The description of each topic will be suitable for both design and engineering students and industrial professionals.

- A wide coverage of topics from interaction design and artificial intelligence covering both theory and practice
- Description of case studies on how to develop and evaluate intelligent user interfaces in different domains
- Suitable for a semester (6 months)-long course with a concise description of relevant topics and a self-assessment multiple-choice questionnaire
- Accompanying lecture notes and video lectures for each topic, which will be useful for self-study and online courses
- Coverage of nontraditional, yet important topics of standardization and statistical hypothesis testing in the context of interaction design

The book does not intend to replace existing textbooks on AI and so has not discussed detailed mathematical derivations of AI algorithms or models, rather gives a high-level overview of latest models like Vision Transformers or Diffusion Models with respect to their use cases for interaction design. However, appendix 4 lists a set of basic math topics that an instructor should master to teach the book. The lecture notes https://cambum.net/PB/lecture-notes.php also provide details on individual topics.

Chapter 2

Human factors

2.1 INTRODUCTION

Why do we need to study human factors? Well, there is a simple and common-sense answer: we are making things for humans. In more technical terms, the success of any product, service or system will depend on user acceptability. User acceptability will be decided based on productivity, safety, enjoyability and similar key performance indicators (KPIs). So, we will first try to learn what makes us human, what is special about us. We will not go into philosophy but stay technical. We will also not try to summarize the whole gamut of medical science or psychology in a single chapter. We will rather cherry-pick a few aspects of human perception, cognition and motor action that are relevant for interaction design or investigated by interaction designers.

We will start with understanding the basic working principles of eyes and visual perception. As a member of the family of higher-order primates, 80% of our sensory information is captured through vision. Next, we will briefly discuss auditory and perceptual sensation. Perception will be followed by cognition and various attempts at cognitive modelling. After cognition, we will discuss motor action – mainly rapid aiming movements. Finally, we will learn about various efforts in user modelling and their potential applications.

2.2 HOW WE SEE

Our eyes work like a variable resolution camera. The iris and pupil work like an aperture control, and an inverted image is created on the retina. The macula region of the retina has the highest resolution and has red and green colour-sensitive cone cells. Outside the macula region, the resolution reduces. The rod cells are arranged slightly away from the macula, and hence we can often see better in dusk or dark conditions by looking slightly off focus. Behind the retina, the retinal ganglion cells take the visual signal to the optical cortex, which is located at the backside of the head (Figure 2.1).

DOI: 10.1201/9781003646716-2

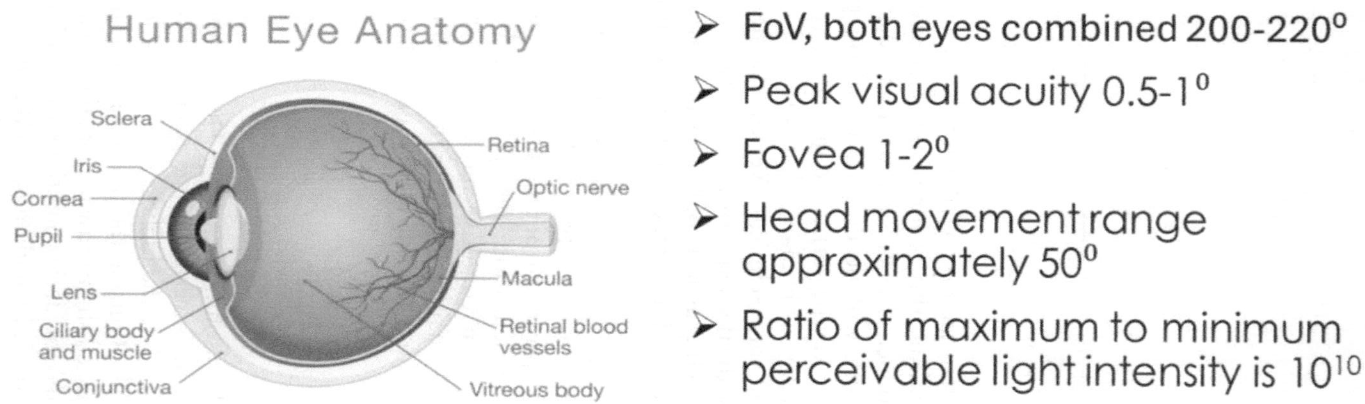

Figure 2.1 Detailed anatomical illustration of the human eye, labelling its various components such as the cornea, iris, pupil, lens, retina, optic nerve and macula. The visual perception metrics of human vision show a wide combined field of view (200–220 degrees) for situational awareness, yet with peak acuity limited to a tiny 0.5–1 degrees foveal region. This fovea, spanning 1–2 degrees, is critical for detailed perception, while our visual system's incredible dynamic range (10^{10}) allows us to see across vast differences in light intensity. This is achieved through a combination of pupil size adjustment, photoreceptor adaptation and neural processing. Head movements further expand our visual exploration by approximately 50 degrees.

The optical cortex has different processing regions: V1, V2, V3, V4 and V5. The V1 and V2 regions have neurons sensitive to shape and orientation features of visual stimuli and work similarly to feature extractors. The V3, V4 and V5 regions take care of eye gaze movement, semantic interpretation and other higher-order functions (Luck et al. 1997). On both conceptual and physiological levels, visual processing can be explained through what and where pathways. The what pathway interprets visual stimuli, while the where pathway creates a salience map localizing individual objects within the visual field (Tovee 2008) (Figure 2.2).

Now, how does this information help us to develop user interface or user interaction? The manifestation of visual processing is observed through two types of visual search: serial and parallel search. If we have similar-looking or all dissimilar objects within our visual field, we need to scan each item to find the desired target. The visual search time will be proportional to the number of objects in our visual field. However, if one object or area within our visual field is distinct, that part or object pops out thanks to the feature-level processing at the optical cortex. The emergency button in a physical interface or Home button in a digital interface can leverage this pop-out effect of parallel search. The number of elements in the visual field should be less to reduce serial search time. We will later read about the glass cockpit that replaces numerous analog dials with one or three multifunction displays showing relevant mission-specific information. Our eyes have variable resolution, and we see best in the red–green region of the spectrum, which can be exploited in interface design. However, it should also be noted that red–green colour blindness is most common, and if the end user group contains colour-blind users, other colour combinations should be exploited beyond the red–green combination.

2.3 HOW WE HEAR

Auditory perception works in a similar way to visual signal processing. The pinna of our ear works as an antenna to focus acoustic waves inside the ear canal. It may be noted that acoustic waves are transmitted both through the ear canal and bones near the ear. We can hear acoustic waves between 20 Hz and 20 kHz. The eardrum functions similarly to the retina and vibrates based on sound waves. The vibration is transmitted through the bones of the middle ear region; note that the three bones in the middle ear are the smallest bones in the human body. The cochlea and vestibular apparatus transmit the sound waves to the brain for further processing. The spiral configuration of the cochlea allows for differing frequencies to stimulate specific areas along the spiral, which results in a tonotopic map that enables humans to perceive various frequencies of sound. The vestibular system also contributes to the sensation of proprioception, orientation

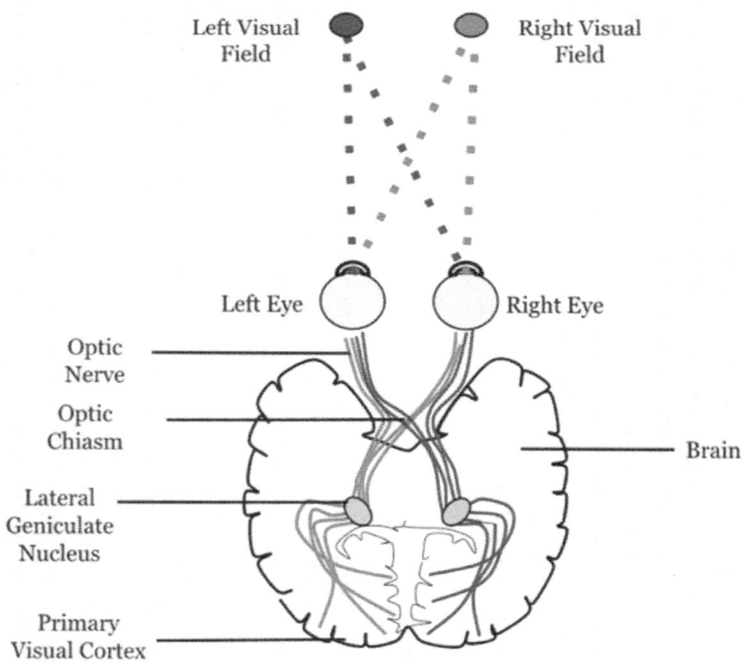

- Visual Pathway
    - Where and What Pathway
    - M and P pathway
- V1 – Primaryvisual cortex
    - Feature extraction
    - Colour and orientation feature
    - Mapping of retina on surface of cortex
- V2 region
    - Visual orientation map
    - Colour map
    - Disparity map
- V4 region
    - Colour recognition
    - Object discrimination
- V3 and V5 regions
- Motion
- Stereoscopic vision
- Visual guidance and scanning

Figure 2.2 Visual pathways showing how information from the left and right visual fields is processed by each eye and relayed through the optic nerve, optic chiasm, lateral geniculate nucleus and ultimately to the primary visual cortex in the brain. The visual pathway of the brain explains how visual data is processed through distinct cortical regions. The ‘where and what pathway’ branches from the M and P pathways to V1, the primary visual cortex, responsible for fundamental feature extraction like colour and orientation, and mapping the retina. Subsequent areas like V2 (visual orientation, colour and disparity maps) and V4 (colour recognition and object discrimination) handle increasingly complex visual attributes, while V3 and V5 are specialized for motion detection, contributing to stereoscopic vision and overall visual guidance and scanning.

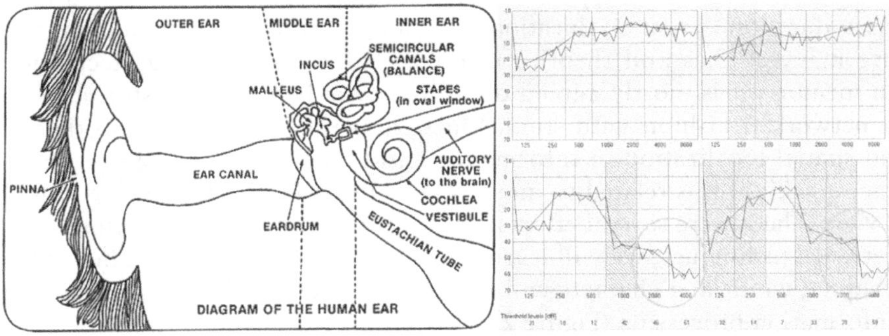

Figure 2.3 (Left) The anatomy of the human ear, detailing the outer, middle and inner ear structures, including the pinna, ear canal, eardrum, malleus, incus, stapes, semicircular canals, cochlea and auditory nerve. (Right) Audiograms for both ears, showing hearing thresholds in decibels across a range of frequencies. The circles on the audiogram indicate a significant drop in hearing sensitivity at higher frequencies (around 2 kHz and 4 kHz), suggesting potential hearing loss in those ranges.

and acceleration of the head in any direction with associated compensation in eye movement and posture. The audiogram measures our sensitivity to different frequencies of sound in terms of measuring the minimum amplitude required to hear a beep of a specific frequency (Figure 2.3). As people get older, they gradually lose the ability to hear sounds of higher frequency and also lose the ability to distinguish across different frequencies. People with conductive hearing loss only require higher amplitude sound to listen properly, but with sensorineural hearing impairment, people often hear slurred voices and background music, or sound may deteriorate the hearing experience (Moore 1997).

2.4 HOW WE COGNIZE

After perception, we will look into the basic processing of information inside the brain, or cognition. The brain can be divided into a set of finite (often 24) functional regions, although multiple regions often work together to undertake tasks. As we shall elaborate in the next chapter, one primary aim of artificial intelligence was and still is to model the human information processing pipeline. Cognitive modelling efforts and associated psychological studies were instrumental to understanding human cognition as well as standard clinical or physiological studies. For interaction design, we can concentrate on the aspect of information retrieval and ways to facilitate that in an interface. Our memory can be categorized into three types based on information retention duration: sensory memory (retention duration 250 msec), short-term memory (retention duration approximately 20 seconds without rehearsal and approximately 8 minutes with rehearsal) and long-term duration (retention duration is 2 minutes to lifetime). Although there are many studies and theories involving information storage techniques in memory, it is mostly agreed upon by researchers that we store facts in a networked fashion with similar facts networked together, and there is no spatial localization of information inside the brain. We can retrieve a fact quickly if it is related to many other similar facts (Hampson & Moris 1996). There were several efforts on modelling and predicting information retrieval techniques; we will learn about cognitive architecture later in this chapter, while later chapters will furnish more details on visual and textual information processing (Figure 2.4).

2.5 HOW WE ACT

We act in various ways by moving our various body parts such as walking, running, nodding heads, hand gesturing and so on. In the field of user interaction, we concentrate on a particular type of movement known as rapid aiming movement. We will discuss haptic sensation in Chapter 4. Analysis of rapid aiming movement has a rich history starting from the last decade of the 18th century, when Woodworth (Eliott et al. 2001) was inspired by the speed and accuracy of the hammering activity of construction workers and concluded that with visual feedback people make more errors with higher speed of execution while without visual feedback, error does not depend on accuracy. The finding leads to two phases of rapid aiming movement: the ballistic phase that does not depend on visual feedback and the homing phase that depends on visual feedback. In 1954, Paul M Fitts published a seminal paper where he predicted the task completion time as a function of distance to target and width of target (Figure 2.5). In 1979, Stuart Card used Fitts' law to predict the mouse as the most efficient pointing device compared to the joystick and head movement tracker based on which Xerox Parc and later other companies started to market the computer mouse. Later, Fitts' law task became part of the ISO 9241 standard and is presently used to compare any pointing device. Besides numerous terrestrial applications, Fitts' Law study was also conducted at the International Space Station ([Holden 2023; Biswas 2026]). Fitts' Law found to work in microgravity environment as well in terms of more than 0.9 correlation between movement time and 2-Dimensional Index of Difficulty (Biswas 2016) while both studies (Holden 2023; Biswas 2026) did not find any significant difference among pointing and selection times with touchscreen tablet between microgravity and ground environments after crews settled down at International Space Station.

TIME SCALE OF HUMAN ACTION		
SCALE (sec)	SYSTEM	STRATUM
10^7 10^6 10^5		SOCIAL
10^4 10^3 10^2	Task Task Task	RATIONAL
10^1 10^0 10^{-1}	Unit Task Operations Deliberate Act	COGNITIVE
10^{-2} 10^{-3} 10^{-4}	Neural Circuit Neuron Organelle	BIOLOGICAL

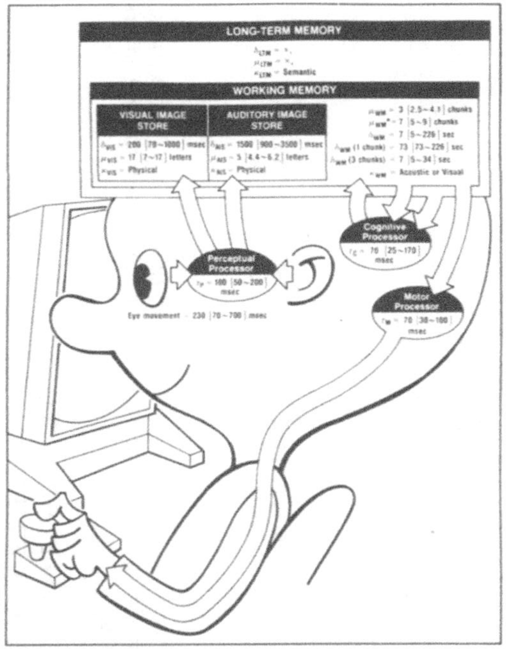

Figure 2.4 (Left) Time scale of human action, categorizing activities based on the duration of execution: biological, cognitive, rational and social. This table helps understand how different levels of human behaviour, from neural circuits to social interactions, operate within distinct temporal frameworks (Newell 1990). (Right) The Model Human Processor, outlining the stages of human information processing from perception through cognitive processing to motor response, along with associated memory stores and their characteristics. It provides quantitative estimates for processing times and memory capacities, such as τP for the perceptual processor (100–200 msec) and μWM for working memory capacity (3–5 chunks). Together, these images offer a foundational understanding of human information processing and the temporal organization of human actions (Card et al. 1983).

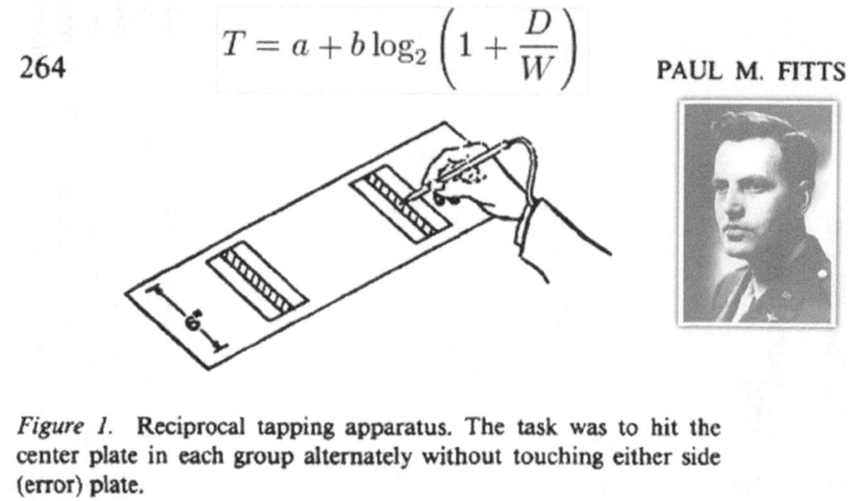

Figure 2.5 Fitts' law formula: $T = a + b \log_2 (1 + \text{Distance/Width})$, where T is the time to execute rapid aiming movement, and a and b are coefficients that are empirically measured.

2.6 USER MODEL: AN EFFORT TO PUT THINGS TOGETHER

A user model can be defined as a machine-readable representation of a user, while a user profile can be defined as an instantiation of the user model. One of the first and pioneering efforts on user modelling was the Model Human Processor (MHP) proposed in the book *The Psychology of Human Computer Interaction* (Card et al. 1983). The MHP described all human–machine interaction as a cycle of perception, cognition and motor action. Later the MHP was further simplified into the GOMS (goal, operator, method, selection) model. The GOMS model works through decomposing interaction tasks into smaller subtasks. It also defines a set of operators with estimated time of completion. Each subtask can be completed by multiple methods and for estimating task completion time, one of the possible methods needs to be selected and executed by a sequence of operations. The KLM GOMS (keystroke-level model) version further simplified the GOMS model by defining only six possible operators, while the CPM GOMS model exploited the parallelism or simultaneous execution of tasks (John & Kieras 1996). The GOMS model can be used to estimate only expert, error-free performance. The NGOMSL model (Kieras 1994) added a natural language processing unit to easily interpret goals and operators from a natural language task description. An example of the execution of the GOMS model for a task involving opening an application on a Windows computer may be as follows:

- Goal
  - Open a folder
- Operator
  - Move mouse
  - Click mouse
  - Press <enter>
- Methods
  - Double click on the icon
  - Select the icon and press <enter>
  - Right click on the icon, select <open> from the pop-up menu

While the GOMS model offered a simple way of modelling human–machine interaction, the cognitive architectures aimed to exploit the cognitive models introduced in the previous section and ideally try to model all cognitive and later perceptual and motor activities of users. The SOAR cognitive architecture only used symbolic processing in the form of representing knowledge through propositional logical statements and rules. Later, the ACT-R model also involved a connectionist model to represent knowledge and predict information retrieval time (Anderson & Lebiere 1998). The EPIC model (Kieras & Meyer 1990) modelled perceptual and motor processing by incorporating Hick's law and Fitts' law, and later, the ACT-R model was changed to the ACT-R/PM model, adding perceptual and motor modules. The cognitive architectures often require extensive parameter tuning and are mainly used to explain interaction rather than predict performance. While GOMS and cognitive architecture-based user models are not developed for a specific application or device, there are numerous examples of application-specific models to predict human–machine interaction or users' intentions. The inclusive user model (Biswas & Langdon 2012), developed at the University of Cambridge, followed the notion of the Model Human Processor and calibrated multiple machine learning models to predict fixation duration, eye gaze movement, cursor trajectory, pointing time and audio distortion for persons with certain visual, hearing and motor impairments. The model was used to design an inclusive user interface by simulating the effect of visual, hearing and motor impairments on interface prototypes (Figure 2.6). The model used a backpropagation neural network to predict fixation duration from colour and shape similarities (Belongie et al. 2002) of stimuli involving users with colour blindness and retinopathy (Biswas & Robinson 2009). It also predicted movement time for four different input devices (mouse, trackball, touchscreen and stylus) from grip strength, active range of motion of wrist, size and distance to target involving users with both hypokinetic (spasm) and hyperkinetic (tremor) motor impairment (Biswas & Longdon 2012). The system also implemented a frequency smearing effect to simulate hearing perception of people with hearing impairment (Biswas & Longdon 2013).

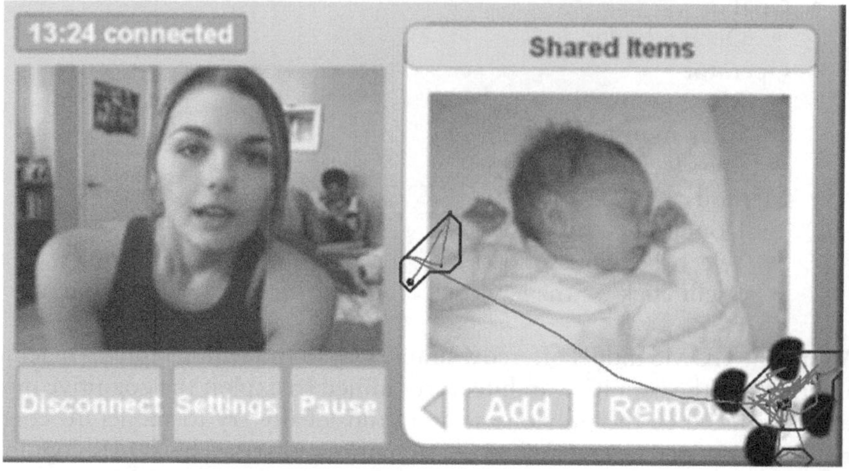

Figure 2.6 Simulation of vision of a user with colour blindness, dry macular degeneration and cursor trace of a user with spasm.

We shall discuss interface personalization and standardization efforts on user modelling in Chapter 11.

2.7 QUICK FACTS

- **Visual search**: A visual search can be either serial or parallel. When different objects in the visual field look similar, the search process is serial, and search time is proportional to the number of objects in visual field. If a particular object or region in the visual field is more distinctive than others, it 'pops out' from the surrounding and the search time is not proportional to the number of objects in visual field.
- **Visual pathway**: The visual pathway is the processing pipeline of visual stimuli in the optical cortex. There are two visual pathways: the what and the where. The what pathway decodes the semantic information of the stimuli. The where pathway provides spatial localization of a particular stimulus within the visual field.
- **Visual angle**: The visual angle of a stimulus is calculated by the arc-tangent (tan–1) of the ratio of its dimension and distance from the viewpoint (eye).
- **Colour blindness**: Colour blindness is a type of visual impairment that impedes the ability to distinguish between colours. The most common type of colour blindness is the reduced ability to distinguish between red and green colours.
- **Audiogram**: An audiogram is a graph that plots the minimum audible amplitude for different frequencies of sound for each ear.

2.8 SUMMARY

This chapter aimed to give a quick overview of relevant human factors related to user interface and interaction design relevant for electronic screens. We discussed topics like visual search, visual angle calculation, cognitive architecture, Fitts' law and different types of haptic devices, followed by concepts of user modelling approaches like Model Human Processor and GOMS. We did not discuss details of ergonomics such as body posture or safety guidelines for physical interface design, but readers may check out references for more information. Now, it will be interesting to note how we design intelligent systems. For a long time, even from the medieval period, one of the primary aims of automation or artificial intelligence was to develop human-like systems. It was based on our own understanding of ourselves; artificial intelligence evolves from symbolic logic to the Nobel Prize for AlphaFold and the backpropagation algorithm and the Turing Award for reinforcement learning. The next chapter will introduce basic concepts of artificial intelligence.

Chapter 3

Basics of AI/ML

3.1 INTRODUCTION

From human intelligence, this chapter will introduce basic concepts of artificial intelligence (AI). While AI is a common buzzword everywhere now, the topic of AI was not always popular – it suffered many ups and downs in terms of application, prospective danger and subsequent academic or industrial interest. This chapter will start with a brief history of the concept of automation, robots and artificial intelligence. While AI can be described or explained in different ways, we will classify different AI models in terms of underlying knowledge representation. As our final goal will be to develop an intelligent user interface or interaction, different types of knowledge representation will help us decide appropriate encoding techniques for incorporating AI in user interface and user experience (UI/UX) design. Among different types of machine learning (ML) models, we will briefly introduce linear regression and neural networks as examples of supervised ML models, the basic context of cluster analysis or unsupervised machine learning and using reinforcement learning for motion planning of autonomous agents. The chapter aims to provide a basic understanding of concepts instead of detailed mathematical foundations, which may be found in the cited lecture notes.

3.2 HISTORY OF AUTOMATION AND ARTIFICIAL INTELLIGENCE

The quest for automation and automated instruments dates back to ancient times. The famous author of the *Iliad* and *Odyssey*, Homer, wrote about wheeled servants and female assistants endowed with sense, speech and strength for the Greek god Hephaestus (Mayor 2018). From the 3rd to 1st century BCE, the Alexandrian school inventors developed many automated systems such as the automatic organ pipe and the mobile shrine of Dionysus, mainly through mechanical and hydraulic means. Outside

DOI: 10.1201/9781003646716-3

Europe, similar automated systems were reported in the Middle East, Egypt, China and India. In Ancient Egypt, statues with moving heads played important roles in religious ceremonies (16th–15th century BCE). The Buddhist scholar Daoxuan (596–667 AD) described humanoid automata crafted from metals that recite sacred texts in a cloister that housed a fabulous clock (Truitt 2015). The Indian Lokapannatti, a collection of cycles and lore produced in the 11th or 12th centuries AD, tells the story of how an army of automated soldiers (*bhuta vahana yanta* or 'Spirit movement machines') was crafted to protect the relics of Buddha in a secret stupa. Chinese carpenter Lu Ban and the philosopher Mozi described mechanical imitations of animals and demons. The implications of humanoid automatons were discussed in *Liezi* (4th century CE), a compilation of Daoist texts. *Samarangana Sutradhara*, a Sanskrit treatise by Bhoja (11th century), includes a chapter about the construction of mechanical contrivances (automata), including mechanical bees and birds, fountains shaped like humans and animals and male and female dolls that refilled oil lamps, danced, played instruments and re-enacted scenes from Hindu mythology. During the 13th century, Ismail al-Jazari, best known for writing *The Book of Knowledge of Ingenious Mechanical Devices*, created automated moving peacocks driven by hydropower and also invented the earliest known automatic gates, which were driven by hydropower, and created automatic doors as part of one of his elaborate water clocks. One of al-Jazari's humanoid automata was a waitress that could serve water, tea or drinks. The word *automaton* was first defined by the French polymath François Rabelais, born in the late 14th century, to denote a machine with a self-contained principle of motion in his book series called Gargantua (Truitt 2015). Considering the present craze with humanoids, the first oracular head is credited to Pope Sylvester II, earlier Gerbert de Aurillac. With further improvement and higher precision of mechanical devices as well as the requirement of mathematical calculations in daily life, the quest for automata moved to developing mechanical calculators. During the mid-16th century, a French teenage boy, Blaise Pascal, invented mechanical calculators to help his father in tax calculation. Although Pascal later gained fame as a mathematician, physicist and perhaps also the inventor of the public transport system, the Pascaline (also known as the arithmetic machine or Pascal's calculator) paved the way for development of modern computers. Jumping a few more centuries, the Second World War further emphasized the requirement for automated machines for large-scale calculation, mainly for code-breaking and trajectory prediction of airborne objects. The word *artificial intelligence* was first popularized at a workshop held on the campus of Dartmouth College in 1956. Organizers of the workshop – Claude Shannon, John McCarthy, Nathaniel Rochester and Marvin Minsky – and many attendees became the leaders of AI research for decades.

3.3 SYMBOLIC VERSUS CONNECTIONIST APPROACH

The early work on AI attempted to represent knowledge in the form of propositional logic statements, logical operators and rules (Russell & Norvig 2009). Rules were represented as if–then statements such as '*If* it rains and the time of day is morning, *then* there will be a traffic jam'. In an abstract way, '*If* Fact A *and* Fact B, *then* Fact C'. Following are examples of facts and rules from the CLIPS (C Language Integrated Production System) expert system.

```
(defrule save_baby_1 (declare (priority 99)
;if
(the baby is crawling across the road)
;and
(a Mack Truck is approaching it)
=> ;then
(assert (get the baby out of danger)))
```

Following is an example of fact/knowledge represented as a statement:

```
(class (teacher "Martha Jones") (#-students 30) (Room "37A"))
```

There was a plethora of systems known as expert systems that proposed the early version of conversational AI agents for different purposes. The Dendral system (DENDRitic ALgorithm) was the first expert system; it was developed in the 1960s to determine chemical composition from mass spectrogram data. It used a series of heuristics and subprograms to determine if the chemical compound generated was plausible given the data. The user could steer the direction that DENDRAL took next by suggesting what chemical elements should be added or removed from the generated hypothesis compound. The ELIZA system, created by Joseph Weizenbaum (1966) at MIT, was one of the first chatterbots and one of the first programs capable of attempting the Turing test. The most famous script, DOCTOR, simulated a Rogerian psychotherapist (in particular, Carl Rogers, who was well-known for simply parroting back at patients what they had just said) and used rules dictated in the script to respond with non-directional questions to user inputs. Implemented in the early 1970s, the MYCIN (Shortliffe & Buchanan 1975) is perhaps the most recognized and cited expert system. MYCIN was developed over five or six years in the early 1970s at Stanford University. It was written in Lisp as the doctoral dissertation of Edward Shortliffe under the direction of Bruce G Buchanan, Stanley N Cohen and others. MYCIN operated using a simple inference engine and a knowledge base of approximately 600 rules. MYCIN received an acceptability rating of 65% on treatment plans from a panel of eight independent specialists, although it was never actually used in practice. The MYCIN system and its potential use case raised ethical questions about the use of AI and the legal implications of the failure of AI, which are still relevant today in the context of autonomous driving and generative AI. It may be noted that the expert systems of the 1960s to 1970s relied on logical operators (AND, OR, XOR) for reasoning. In 1954, Warren McCullough and Warren Pitts published a paper on implementing logic operators through neuron-like structures. Electronic components operate and communicate at a much higher speed through the passing of electrons, while neurons in the human brain operate and communicate via the passage of synaptic fluid and related chemical reactions. Yet there are pattern recognition and information retrieval activities that even a human child can undertake much faster than a computer. Scientists aimed to mimic this interconnected network structure of neurons in digital information processing. In the next section, we will discuss the basic architecture of a feedforward artificial neural network. Chapter 5 on computer vision introduces convolutional neural networks (CNNs), and Chapter 6 on information retrieval covers recurrent neural networks (RNNs).

3.4 ARTIFICIAL NEURAL NETWORKS AND MACHINE LEARNING

In this section, we introduce the backpropagation algorithm for training artificial neural networks. However, before discussing backpropagation, we will learn a technique known as gradient descent in the context of a much simpler problem of fitting a straight line through a set of non-collinear points. It may also be noted here a change in knowledge representation for learning an AI model; whereas in the case of expert systems, knowledge was represented as propositional logic statements, in the case of linear regression or neural networks, knowledge will be encoded as a numeric data table.

Let us consider that the mark obtained by a student will depend on class attendance, and we are trying to predict the mark obtained by a student as a function of number of classes attended by that student. We can collect historical records on students' marks and attendance and make a graph by plotting class attendance on the x-axis and mark obtained on the y-axis, with each point representing a student. Now the simplest model will be a straight line with the equation $Y = aX + b$, where Y represents the mark obtained, X represents the number of classes attended, and a and b are model parameters that we need to empirically calculate. For a straight line, a is the slope and b is the y-axis intercept. For a neural network, as we will shortly find out, there will be a lot more model parameters called weights. There are multiple methods to find *a* and *b* from a given dataset. We will use the gradient descent method; an alternative is the Newton–Raphson method. In gradient descent, we will fit a random model (straight line in this case), calculate the error in prediction and then minimize the error by descending down the gradient of the error (hence the name). While there are many online and offline resources available on the full derivation of gradient descent, readers may presently only note that the algorithm iterates through correcting the initial set of random values of model parameters by changing them through the gradient of the error with respect to model parameters until either the error in prediction or change in model parameters between two successive iterations is below an acceptable threshold.

A single neuron in a neural network takes a set of input and calculates a weighted average of the input (like linear regression) and then limits the output through an activation function. The activation function is different based on the type of network; it can be a Sigmoid function for a shallow artificial network, or it can be a ReLU function for a deep neural network to make sure the whole neural network is taking part in the learning process. A single neuron cannot model complex functions; hence, an ensemble of neurons is connected to make a neural network. The neurons are connected through edges, with each edge assigned to a weight, which serves as the model parameter for a neural network. The learning process of a neural network involves adjusting the weights so that a given set of input can produce a desired output. The backpropagation algorithm is similar to the gradient descent algorithm. It initializes random weights and iterates until either the error in prediction or change in model parameters between two successive iteration is below an acceptable threshold. Each iteration consists of two passes. On the first pass, input data is simultaneously fed to the input layer of neurons, and based on present values of weight, output is calculated. In the second pass, the weights are changed by the gradient of the error (partial differentiation with respect to weights). As the gradients are calculated, the weights are changed from the output layer to the input layer. The name *backpropagation* stands for the process of propagating the change in weights with respect to the error gradient in a backward direction (output to input layer). It is often advantageous to get a neural network predicting the desired output quickly, learning through a smaller number of iterations. Hence, in the learning process, a new parameter is introduced called the *learning rate* that controls the amplitude of change in the weights. In deep learning neural networks, the learning parameter is also adjusted based on the sign (plus or minus) and value of the gradient (Goodfellow et al. 2014). It may be worthy to mention here that one of the co-inventors of the backpropagation algorithm, Prof Geoffrey Hinton, was awarded the Turing Award in 2018 and the Nobel Prize in Physics in 2024 (Figures 3.1 and 3.2).

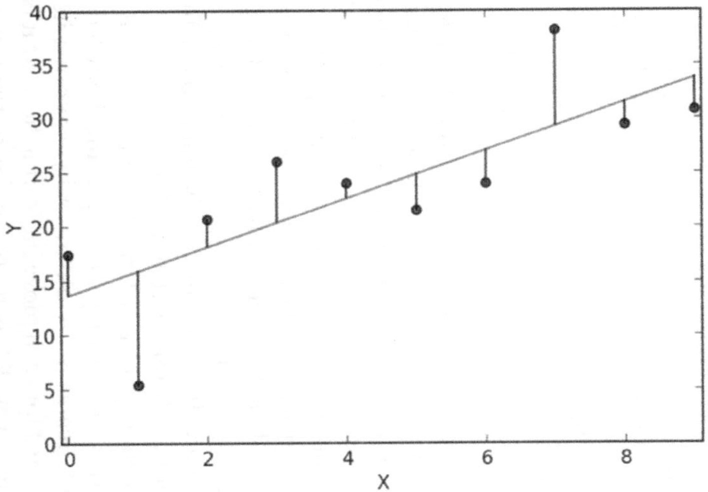

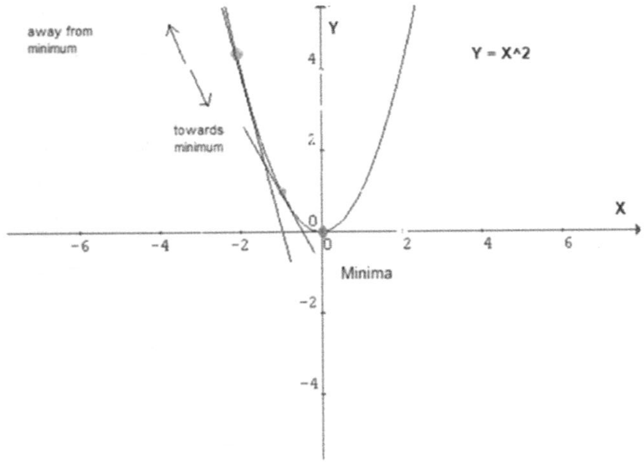

Figure 3.1 (Left) Errors in linear regression and graphical representation of iterations of gradient descent. (Right) Plots of the mean square error and gradients.

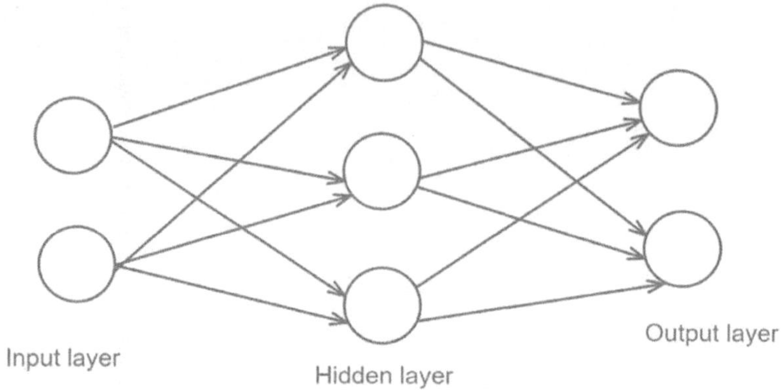

Figure 3.2 Basic structure of a feedforward neural network showing input, hidden and output layers.

3.5 DIFFERENT TYPES OF MACHINE LEARNING

In the earlier example, we considered a case of predicting students' marks from their attendance record. As the model learns through a set of examples, or supervision, the learning technique is known as supervised learning. When we predict one among a finite state of probable outcomes or classes (in our case, students' grades), the underlying technique is known as classification. When we predict a parameter on a continuous scale (for example, students' marks), it is known as regression. In Chapter 9, we will use supervised learning techniques like linear regression for coordinate space conversion in augmented reality-based human–robot interfaces, while an artificial neural network serves a similar purpose in Chapter 10 in the context of an automotive user interface.

However, what if instead of prediction, we wanted to check if students' marks really depend on attendance? A correlation coefficient can provide an indication if there is a linear relationship, but a more interesting observation may be if students with similar marks also have similar class attendance. Clustering or unsupervised learning does not predict one parameter from others; rather, it groups or clusters data based on a similarity index. Like the previous case, data or observations are encoded as a numeric data table, but unlike the previous case, we do not try to predict a specific column. We rather calculate distances among the data points and group them together based on mutual distances (Figure 3.3).

Reinforcement learning (RL) is different from supervised and unsupervised learning in terms of learning technique and knowledge representation. In RL, an agent creates its own dataset by exploring its environment while it receives a reward for each move indicating how far it is from its goal or destination. In some way, RL is more similar to the way we learn from positive and negative examples in our life. In fact, RL was investigated in

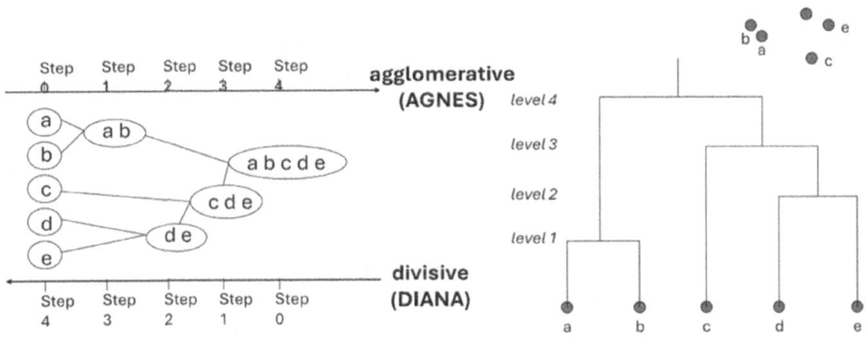

Figure 3.3 (Left) Agglomerative (AGNES) and divisive (DIANA) hierarchical clustering methods, showing how data points are successively merged or split at each step. Agglomerative (AGNES) starts with individual data points (a, b, c, d, e) and merges them into progressively larger clusters, as seen in the steps from 0 to 4, ultimately forming one large cluster. Divisive (DIANA) works in reverse, starting with all data points in one cluster and recursively splitting them until each data point is an individual cluster, as shown in steps 4 down to 0. (Right) A dendrogram, which is a common visual representation of hierarchical clustering results, depicting the clusters and their hierarchy, with labels a, b, c, d and e at level 1 representing individual data points and higher levels showing the merging or splitting of these points into clusters.

detail by psychologists and pedagogists much before computer scientists. Most RL algorithms are explained by representing knowledge in the form of a graph or grid. A graph is a data structure where each state of a system is represented as a node and the change of state is represented as an edge or link connecting two nodes. In the case of path planning problems, we often discretize the physical space in two-dimensional grids or three-dimensional voxels and develop algorithms in terms of moving an agent from one grid (or voxel) to another. As the agent (for example, a rover or drone) moves from one grid, node or voxel to another, it receives a reward with respect to its relative position from the destination. RL algorithms compute long-term rewards (called utility) from instantaneous rewards and set up a policy for an agent at any state. It may be noted that an agent can be a software program as well where a node can represent a specific state of a game or any particular situation. Considering the importance of RL, ACM named Andrew G. Barto and Richard S. Sutton as the recipients of the 2024 ACM A.M. Turing Award for developing the conceptual and algorithmic foundations of reinforcement learning. We shall further discuss RL (Value Iteration Algorithm) in chapter 9 and Inverse Reinforcement Learning in Chapter 11.

3.6 MODEL VALIDATION

How can we evaluate the quality of an ML model? In fact, most of the time we preprocess data and feed it to an ML model treating the model as a black box and checking results. If the result is not acceptable, we change model parameters or change the model itself. In this section, we will look at model validation techniques for each type of ML model.

For supervised learning, we test the model's performance through K-fold cross-validation. K-fold cross-validation involves randomly dividing the dataset into k segments, training the model with (k–1) segments, testing it on the remaining segment and repeating the procedure k times. However, before k-fold cross-validation, we should load-balance the data, which means we should make sure each class or range of values has nearly equal numbers of data points. It involves selecting a class with the lowest number of data points, randomly picking an equal number of data points from other classes, and then running cross-validation on this balanced dataset (Figure 3.4).

The accuracy of classification is measured through the number of false positives (type I error), false negatives (type II error) and functions combining these two metrics. For regression analysis, we can use root mean square error, R^2 value or correlation between predicted and actual values (Figure 3.5).

In machine learning, we try to model an unknown function. If we try to use a model that is less complex than the actual model, we encounter the situation of underfitting. Say, for example, we are trying to model a physical process involving a fourth-order polynomial function (assume we do not know it) through a quadratic function. Underfitting can be detected when both training error and test error do not reduce to an acceptable value and, over multiple iterations, show an asymptotic flat pattern higher than the desired or required threshold. The opposite situation is overfitting, where a model starts to memorize data instead of generalizing over the physical process or event. Overfitting can be detected when the training error keeps reducing over iterations or epochs of learning, but the testing (or validation) error increases (Figure 3.6).

For unsupervised learning, we can use the following cluster validation indices to find the best number of clusters. Numerical measures that are applied to judge various aspects of cluster validity are classified into the following two types:

- **External index**: Used to measure the extent to which cluster labels match externally supplied class labels. Entropy is used as a measure of external index
- **Internal index**: Used to measure the goodness of a clustering structure *without* respect to external information. Sum of squared error (SSE) can be used as a metric of internal index

For reinforcement learning, models or algorithms can be compared in terms of speed of convergence or quality of a policy or solution with respect to

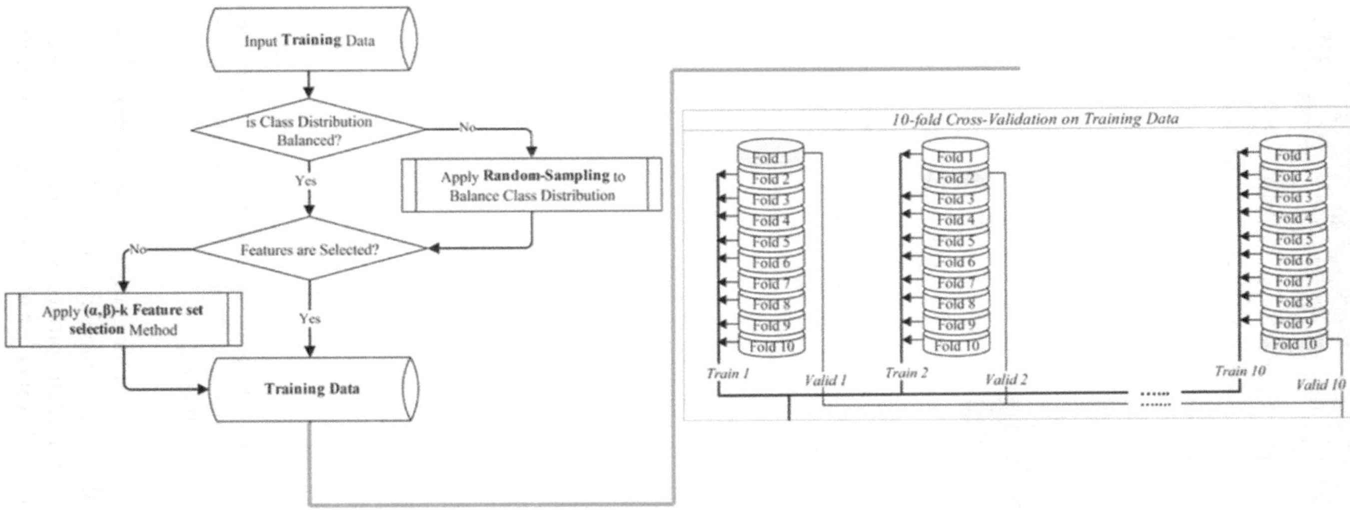

Figure 3.4 A machine learning workflow for training data. It begins by checking if the class distribution of the input training data is balanced; if not, random sampling is applied to balance it. Subsequently, it checks for feature selection: if features are not pre-selected, an (α,β)-k feature set selection method is applied. The processed training data then undergoes tenfold cross-validation, where it is repeatedly split into ten folds, with nine folds used for training and one for validation, iterating ten times to evaluate model performance robustly.

True \ Predicted	A	B	C
A	T_A	E_{AB}	E_{AC}
B	E_{BA}	T_B	E_{BC}
C	E_{CA}	E_{CB}	T_C

- **Accuracy : $(T_A + T_B + T_C)$/# of test samples.**

Figure 3.5 A confusion matrix for a three-class (A, B, C) classification problem. The diagonal elements are correct classifications, while the other cells show wrong classifications.

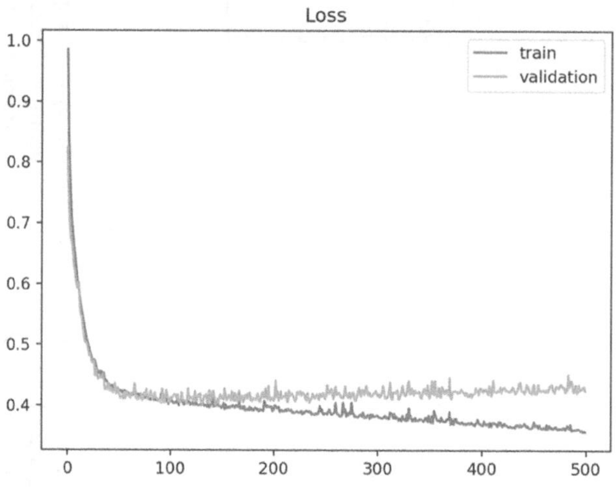

Figure 3.6 Example of overfitting. The graph plots loss on the y-axis and iterations on thex-axis.Overfittingisindicatedwhenthelossstartsincreasingforthevalidationset but decreasing for the training set.

a cost function. For example, algorithms can be compared as the number of iterations, time and space complexity of each iteration for stabilizing utility values or policies for each state of a problem. The cost of reaching a solution in terms of time, distance or any other cost function can be used to compare.

3.7 QUICK FACTS

- **Symbolic AI**: Symbolic AI represents knowledge as propositional logic statements, and the knowledge is processed through firing explicitly encoded rules in propositional logic
- **Linear regression**: Linear regression analysis is used to predict the value of a variable based on the value of another variable. The variable one wants to predict is called the dependent variable. The variable used to predict the other variable's value is called the independent variable (regressor). It may be noted that the word *linear* means the equation is linear in terms of coefficients
- **Neural network**: The human brain can recognize patterns or faces without using explicit conscious rules. A neural network is a machine learning program, or model, that makes decisions in a manner similar to the human brain by using processes that mimic the way biological neurons work together to identify phenomena, weigh options and arrive at conclusions
- **Trees and graphs**: A graph is a data structure G = (V, E) consisting of a set of objects V = {v1, v2, ...} called vertices, and another set E = {e1, e2, ...}, whose elements are called edges. Each edge connects two vertices either in an ordered (directed graph) or unordered manner (undirected graph). A graph G is said to be connected if there is at least one path between every pair of vertices in G. A tree is a connected graph without any circuits
- **Motion planning**: Motion planning aims to find the best path optimizing a criterion to move an agent from one location to another avoiding obstacles. For motion planning in discrete space, the input is often represented as a graph or grid

3.8 SUMMARY

This chapter aimed to give pointers to a few relevant topics of AI and ML. The reader should now be aware of knowledge representation using data tables and graphs. We also discussed the difference among supervised, unsupervised and reinforcement learning. Future chapters will discuss further details on individual algorithms in the context of UI/UX-related applications. For example, Chapter 5 will use the concept of supervised learning and neural networks, while Chapter 9 will dive into further details on motion planning in terms of maintaining a safe distance from the human operator and developing semi-autonomous robotic agents. Armed with the basic concepts of AI and ML, we will next look into alternative or unconventional interaction technology and the challenges in developing and deploying such technology for different interaction scenarios.

Chapter 4

Modalities of interaction

4.1 INTRODUCTION

The previous two chapters were brief overviews of the human information processing system and different knowledge representation techniques in artificial intelligence. This chapter will explore how the two fields work together to exploit human sensory and motor systems to facilitate information processing through different modalities of interaction. We will introduce different modalities of interaction using eye gaze tracking, gesture recognition, speech recognition, AR/VR displays, haptics and spatial audio interfaces. For each of these topics, we will initially learn the definition, then the advantages and disadvantages, followed by implementation principles and technical challenges towards deployment. We will also highlight present and prospective use cases for each modality and the basic concepts of developing a multimodal interactive system.

When we use interactive systems, there is always a primary modality of interaction. In Chapter 1, we learned about the emergence of the mouse as a primary mode of interaction with personal computers in the context of a discussion on Fitts' law. Similarly, we use a touchscreen for interacting with smartphones and physical switches to interact with interactive displays of automotive and industrial instruments and even the SpaceX Dragon capsule. However, there are many occasions where the primary modality of interaction needs to be augmented or replaced by other modalities. Two such occasions are physical and situational impairment – for example, users with different ranges of abilities or an able-bodied user operating a device under certain physical constraints. While a driver is driving a car and trying to operate the instrument panel, or an industrial worker is remotely operating a robot, they may need to 'feel' the object picked up by the robot. All these situations require alternate modalities of interaction. Sometimes we may also need to perceive (see, hear or touch) more information beyond the capabilities of traditional interactive systems and explore uses of immersive media, spatial audio and haptics. This chapter will introduce these new technologies, while a Chapters 9–11 will further elaborate on use cases.

DOI: 10.1201/9781003646716-4

4.2 INPUT MODALITIES

This section will discuss three different alternative modalities of interaction, the standard processing pipeline for each of them and existing technical challenges. We will discuss automatic speech recognition (ASR) systems, gesture recognition systems and eye gaze controlled applications.

Speech recognition systems are presently the most common hands-free input modality for a wide variety of applications from handheld devices to automotive and aircraft cockpits. The basic processing pipeline for speech processing starts with converting acoustic (speech) signals to phonemes and then combining the phonemes into meaningful keywords using a language model. A language model is a statistical model of a language based on the assumption that the probability of the next word in a sequence depends only on a fixed-size window of previous words. As of 2025, it has been superseded by recurrent neural network-based models, which have been superseded by large language models (LLMs). Earlier or safety-critical speech recognition systems (like in cockpits) often limit the vocabulary to a small set of keywords, while consumer electronic systems utilize advancements in natural language processing and LLMs and can understand any spoken voice. Chapter 6 will further explain the natural language processing pipeline. The main challenges with present speech recognition systems include understanding different accents, less spoken languages and interpreting sounds in noisy backgrounds.

Gestures are expressive, meaningful body motions, i.e., physical movements of the fingers, hands, arms, head, face or body with the intent to convey information or interact with the environment. Activity may be defined over a much longer period of time than what is normally considered a gesture; for example, two people meeting in an open area, stopping to talk and then continuing on their way may be considered a recognizable activity. Gestures can be classified into:

- **Semiotic:** To communicate meaningful information
- **Ergotic:** To manipulate the environment
- **Epistemic:** To discover the environment through tactile experience

Gesture recognition systems can be classified into:

- Pen-based gesture recognition
- Tracker-based gesture recognition
    - Instrumented gloves
    - Body suits
- Passive vision-based gesture recognition
    - Head and face gestures
    - Hand and arm gestures
    - Body gestures

The processing pipeline of a gesture recognition system is similar to that of a speech recognition system, with the input being video or a set of trajectories for pen-, touchscreen- or touchpad-based systems. For video input, a computer vision model, commonly using convolutional neural networks (CNNs; discussed in Chapter 5), tracks relevant body parts like the head, wrist, finger or full body. A time-variant machine learning (ML) model tracks body parts, and a supervised ML model, mostly an artificial neural network (ANN), classifies the movement pattern in one of a finite set of gestures. The main challenges with gesture recognition systems are classifying gestures and detecting body parts in different ambient lighting conditions for vision-based systems. Tracking gestures from multiple persons, even if multiple persons can be tracked by a single camera, is one more open challenge.

Research on eye tracking dates to the late 18th century when Louis Émile Javal investigated saccadic movements in a reading task. Eye tracking is the process of measuring either the point of gaze (where one is looking) or the motion of an eye relative to the head. An eye tracker is a device for measuring eye positions and eye movement (Duchowski 2007). Analyzing points of fixation (Olsen 2012) and eye movement is extensively used to investigate:

- Areas of interest in a display
- Reading behaviour
- The affect of the user
- Visual impairment like nystagmus (irregular eye movement)
- Design of billboards, traffic signboards and so on

Gaze control is about affecting computer action by changing the direction of one's gaze (eye movement), blinking or dwelling on an object. Gaze control interfaces are often used for people with severe motor impairment who cannot use other computer peripherals or applications where operators are already occupied with other tasks, like drivers or pilots. Eye tracking devices can be classified into the following types:

- **Eye wearables**: Edmund Huey built the first eye tracker in 1908, which was a contact lens connected with an aluminium pointer. The first generation of eye trackers and recent wearable contact lenses offer eye tracking using mechanical means or using wearable electronics. These are least popular and will not be discussed in detail
- **Infrared-based eye tracking**: This is the most popular and accurate eye tracking presently available. These are separate devices that can be mounted below a screen or can be attached to glasses or virtual reality devices. The device consists of infrared (IR) illuminators and small-sized high-speed cameras. Eye gaze is estimated by tracking pupil rotation or movement of corneal reflection. A calibration

program usually implements a supervised learning algorithm that maps eye gaze in 2D screen or 3D space (Figure 4.1)

- **Appearance-based eye tracking:** This type of eye tracking does not require a separate device or illuminator but takes a webcam or images in the RGB spectrum as input and estimates eye gaze using a deep learning-based computer vision algorithm. As of 2025, appearance-based eye tracking is ten times less accurate compared to IR-based eye tracking but is often useful for distraction detection or gaze control interfaces with limited screen elements. We will discuss one such application in Chapter 9

Presently, the main challenges in eye tracking are locating the pupil or pupil region at high ambient illumination (greater than 3000 lux) and taking care of a wide range of head and eye movement.

4.3 EXTENDED REALITY (XR) DISPLAYS

In 1964, the Uniscope 300 machine had a built-in CRT (cathode ray tube) display. While not a true computer monitor, it was a precursor to CRT monitor technology, and the first resistive touchscreen display was developed by George Samuel Hurst in 1975. The Senorama simulator was patented in 1962 in an effort to develop 3D movies and is often credited as the first immersive system. In the past 60 years, the world has moved from CRT displays to high-definition OLED displays. In this chapter, we will concentrate on extended reality (XR) systems. Broadly, XR systems can be classified into mixed or mixed augmented reality systems and virtual reality systems. The term 'augmented reality' (AR) is attributed to former Boeing researcher Tom Caudell, who is believed to have coined the term in 1990. When the synthetic imagery is rendered in 3D, the system is often known as mixed augmented reality, mixed reality or augmented virtuality. The International Standardization Organization defines mixed and augmented reality systems as a system that uses a mixture of representations of physical world data and virtual world data as its presentation medium. The virtual reality system creates a completely synthetic environment where users can immerse themselves without any vision of surrounding reality. The International Standardization Organization defines virtual reality (VR) as a set of artificial conditions created by computer and dedicated electronic devices that simulate visual images and possibly other sensory information of a user's surroundings with which the user is allowed to interact.

Augmented or mixed augmented reality systems (Dieter & Höllerer 2016) can further be classified as optical and video see-through systems. In optical see-through systems, viewers can see the real world with the naked eyes, while in video see-through systems, one or more outward-looking

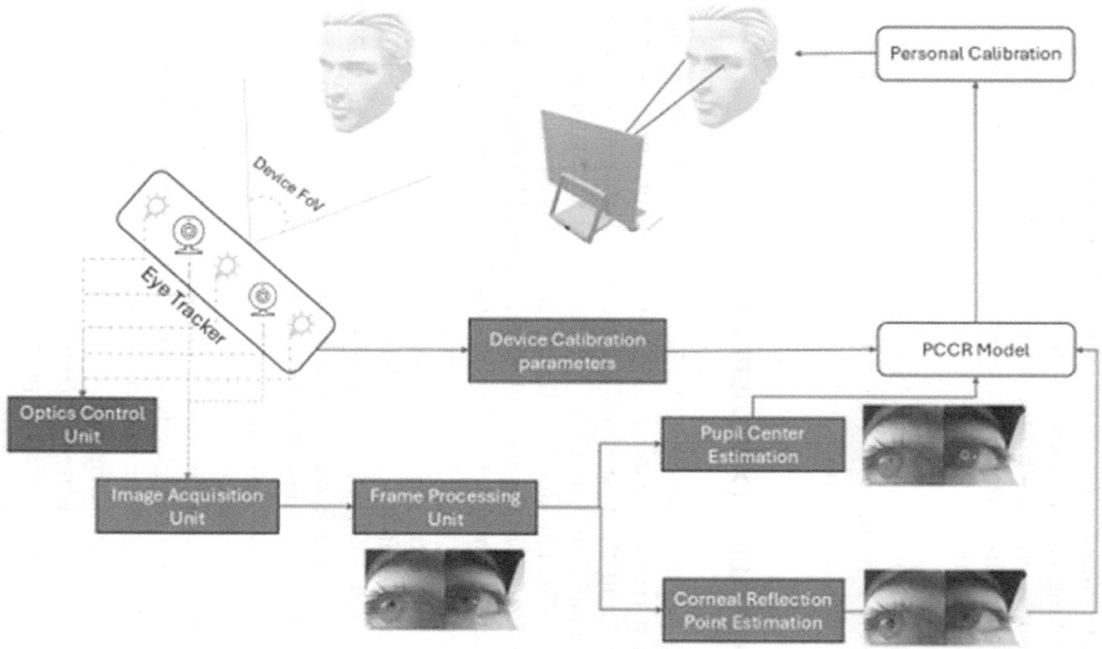

Figure 4.1 The figure shows different parts of an IR based eye tracker. The Optics Control Unit synchronizes cameras and Infrared illuminators by dynamically turning them on and off. The Image Acquisition Unit captures images with predefined settings like exposure time, fps and resolution. The Frame Processing Unit detects face and subsequently eye regions cropped from the image frame. The calibration process sets up individual model based on user and usage context like distance from the screen. The Pupil Centre Corneal Reflection (PCCR) model works by tracking the relative positions of the pupil and a corneal reflection (also known as a glint) to determine where a person is looking.

cameras bring a live video feed inside a headset or handheld display device. The synthetic imagery can also be two-dimensional for handheld tablets, while it can be three-dimensional holograms for mixed reality headsets. We will discuss many such systems in Chapters 9 and 10. Projected augmented reality systems (often known as situated AR) use projectors and computer vision technology to track a physical surface and project synthetic imagery on it. They found applications for projecting brand labels on physical products. Main technical challenges for AR systems include tracking real-world objects in real time, adjusting the brightness of synthetic imagery based on ambient illuminance, and vertical field of view of video see-through displays.

The majority of virtual reality systems (Burdea & Coiffet 2016) are headset based following pioneering work by Ivan Sutherland for his Sword of Damocles headset in 1968 (Figure 4.2). However, there are examples of projected 3D systems and autostereoscopic displays, where users can view 3D content without wearing a headset. As of 2025, the projected 3D systems are not portable or commercially available, and autostereoscopic volumetric displays have less temporal resolution for interactive applications.

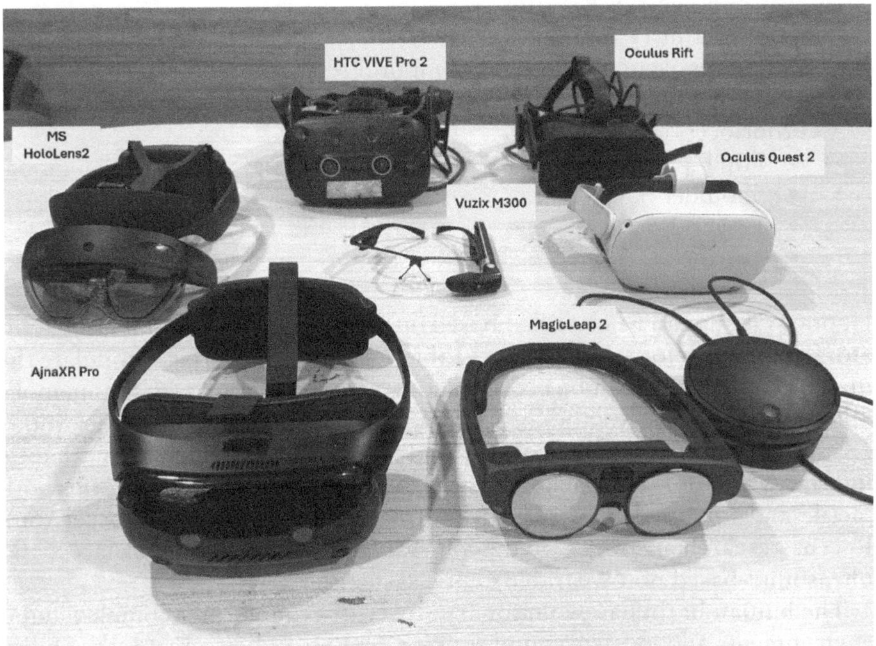

Figure 4.2 Extended reality displays. HTC Vive and Oculus Rift are tethered VR headsets, Oculus (Meta) Quest is a non-tethered VR headset, Microsoft HoloLens and Magic Leap are optical see-through MR headsets, AjnaXR Pro is a video see-through mixed reality headset and Vuzix M300 is an augmented reality headset.

The main challenges for VR systems are simultaneously maintaining both spatial and temporal resolution of virtual displays and providing realistic perceptual feedback.

4.4 HAPTICS AND SPATIAL AUDIO

Haptics is an information processing perceptual system that uses inputs from the receptors embedded in the skin, as well as in muscles, tendons and joints. The haptic system can sense and act on the environment, while vision and audition systems have purely sensory natures. Human haptic perception has the following range and resolution:

- Temperature sensor
  - Nociceptors (–15°C to 45°C)
- Tactile perception
  - Active touch – stiffness >25 N/mm is needed for an object to be perceived as rigid
  - Spatial resolution
    - Fingertips 2.5 mm
    - Palm 11 mm
  - Successiveness threshold (temporal resolution)
  - Skin 5 msec (compared to temporal resolution of eye is 25 msec)
- Proprioception
  - Hip 0.2°
  - Shoulder 0.8°
  - Finger 2.5°
  - Toe 6.1°

Haptic systems can be classified based on type of feedback, number of actuators, maximum force and weight (Table 4.1). Tactile interfaces simulate the direct touch and feel of objects contacting skin using SMAs (shape-memory alloys), pneumatic actuators or vibrotactile elements. Force feedback interfaces exert simulation through force. The majority of the existing haptic interfaces are force feedback interfaces, while a few haptic gloves offer both tactile and force feedback (e.g., Maestro Glove and HaptX Exoskeleton). Recent research also explored temperature-based haptic interfaces and ultrasound-based non-wearable haptic systems.

The human brain has evolutionary structures to interpret complex audio environments and localize sound sources. Spatial audio recreates the ability to precisely locate the audio source in 360-degree space. Head tracking is important to provide accurate spatial audio. Custom head-related transfer function (HRTF) is needed for each individual, but approximate HRTFs will be useful in many situations. A HRTF is a mathematical model that describes how sound changes as it travels from a source to a listener's ear,

Table 4.1 Examples of commercially available haptic devices

Product name	*Type of feedback*	*Number of actuators*	*Weight (g)*
iFeel Mouse	Vibrotactile	1	132
CyberTouch glove	Vibrotactile	6	142
DTSS X10	Temperature	<8	340
WingMan 3D Force joystick	Force	2	NA
PHANTOM Desktop	Force	3	75
PHANTOM 1.5/6.0	Force	6	90–108
Haptic Master Force	Force	3	NA
CyberGrasp Force glove	Force	5	539
Senso Glove	Electromagnetic tactile	5	na
Cynteract	Electromagnetic force	5	na
Maestro	Electromagnetic force	5	590
GoTouchVR	Electromagnetic tactile thimble	1	20
Tactai Touch	Vibrotactile thimble	1	29
Dexmo	Electromagnetic force	5	320
VRgluv	Electromagnetic force	5	NA
Sense Glove DK1	Electromagnetic	5	300
HGlove	DC motor force	3	750
Manus Prime 3 Haptic XR	Integrated Vibrotactile Haptics	5	138
CyberForce arm	Force	8	NA
Gloveone	Electromagnetic	10	NA
AvatarVR	Vibrotactile	10	NA
HaptX	Microfluid force Feedback	30 per hand	1000 (approx.)
UltraHaptics	Ultrasound array of size 50–700 mm, $\pm 30°$ cone centred around top surface of transducer board	16×16 transducer array (approx.)	700

NA, not available.

influenced by the head, torso and pinnae (outer ear). An accurate spatial audio rendering pipeline requires head tracking, a HRTF and model of the enclosed space to simulate reverberation. Commercial head-tracked headphones are often sufficient to create spatial audio. Responsive spatial audio for immersive gaming (https://www.microsoft.com/en-us/garage/profiles/responsive-spatial-audio-for-immersive-gaming/) can be used to add spatial audio to VR environments (Figure 4.3).

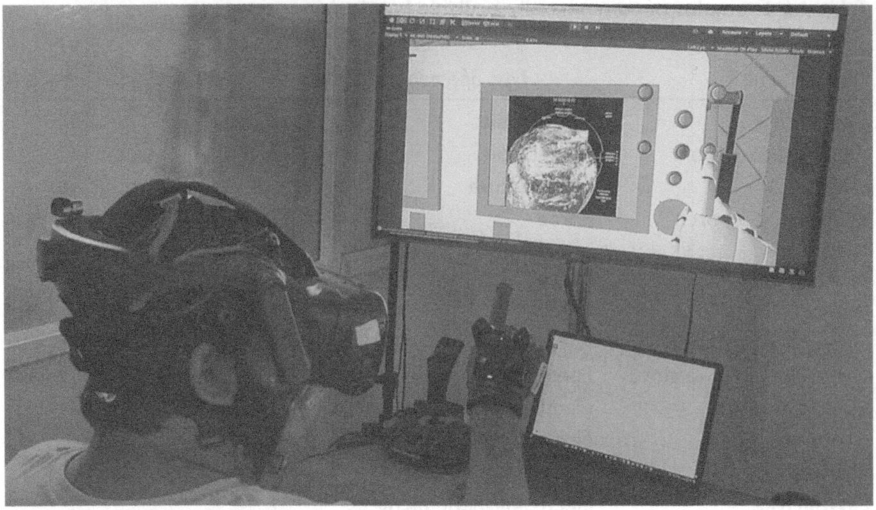

Figure 4.3 A person operating a virtual reality spacecraft simulator with haptic gloves.

4.5 MULTIMODAL INTERACTION

Since we discussed multiple modalities of interactions, in this section we will discuss strategies to utilize multiple modalities together. Different modalities of interaction can be combined based on the following strategies:

- **Time of fusion**: Fusion is classified into pre-, mid- and post-mapping with respect to the time of mapping raw sensor data into recognizable symbols
- **Information to be fused**: Information can be fused at the raw data level, feature level or decision level
- **Type of cooperation among different modalities**: Different modalities can coexist simultaneously or one at a time, or a specific modality can be used for specific information
- **Methods of fusion**: Different modalities can be fused following rules or modelling them as a set of prior probabilities in a Bayesian model, or using filtering techniques on multiple modalities to estimate a probability distribution function

The following example demonstrates a case study of teleoperating a haptic-enabled robotic manipulator from a VR interface. Figure 4.4 depicts a schematic diagram of the proposed system. The modalities are listed as the output and input devices. The XR teleoperation core consists of input selection, a dialog manager, multimodal fission and a state model. The XR teleoperation core receives input from input devices, transfers the state of the XR application and then conveys output to the user through output devices.

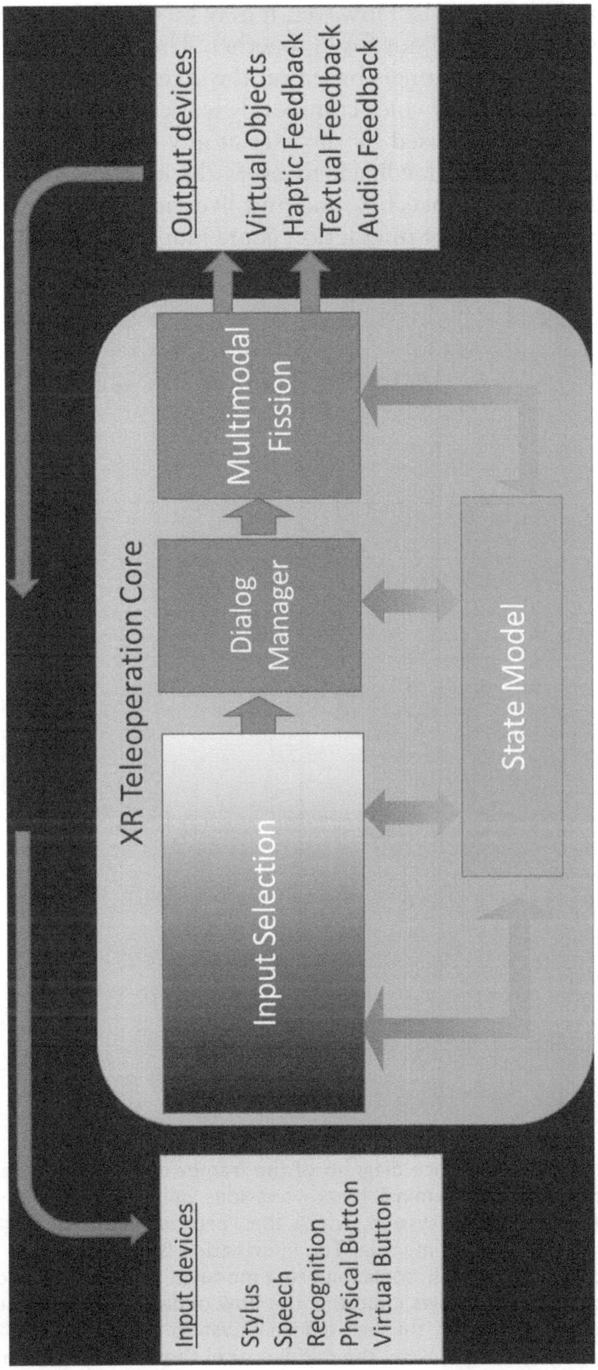

Figure 4.4 The XR-based teleoperation system's architecture.

The user can operate the system using physical and virtual buttons, a 2-DoF stylus and spoken commands. However, it may be noted that the inputs are processed one at a time and not fused together. The input selection module selects one of the input channels based on the state of the operation of the system. Virtual buttons and voice commands work as redundant modalities – any one of them can be used by the user at any time during the interaction. The system provides feedback to users through a set of holograms, haptics, spoken and non-speech sound and live video of the robot's workspace. Based on the phase of interaction (state model), feedback in multiple

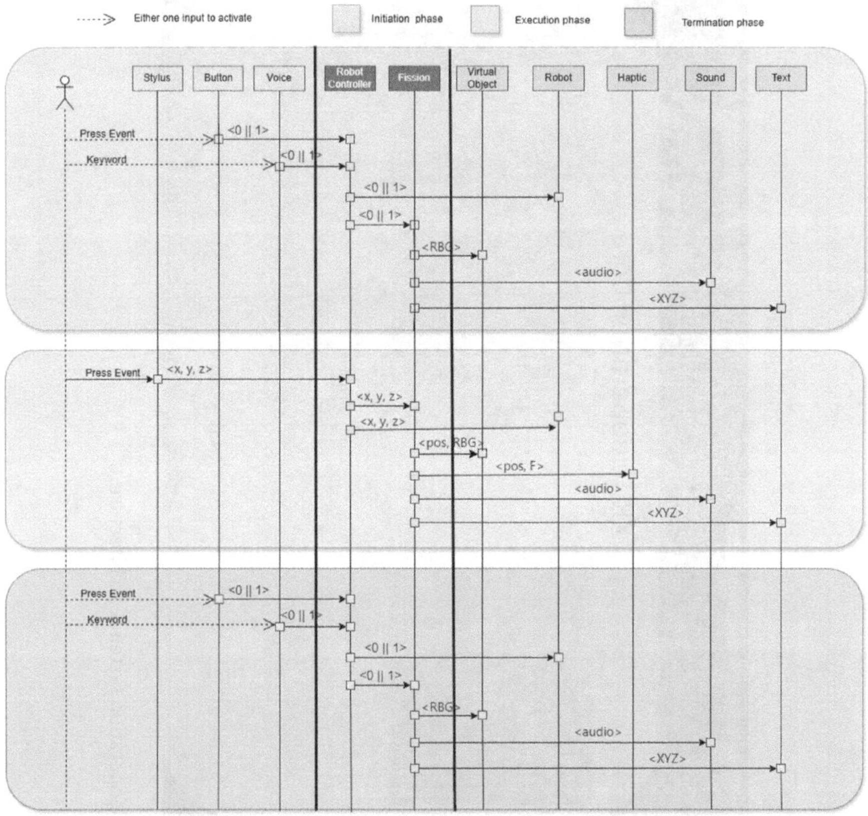

Figure 4.5 Illustrates the sequence diagram of the framework across three phases: initial, execution and terminal. It shows various input modalities (stylus, button, voice) activating the system, which then process inputs through the robot controller and fusion module. The information is sent to virtual objects, the robot, haptic feedback, sound and text modules. Each phase has corresponding actions, with arrows indicating the flow of data and events, such as press events and keywords, through different system components. The top panel shows the initiation phase, the middle panel shows the execution phase and the bottom panel shows the termination phase.

modalities is provided through a multimodal fission module. The fission module monitors the phase of operation and activates appropriate feedback modalities as described in subsequent paragraphs. The system response in Figure 4.5 consists of both actions by the teleoperated robot as well as corresponding changes in the XR interface of the user. Details of the project can be found in Raj et al. (2024).

4.6 QUICK FACTS

- **Eye gaze tracking**: It is a process of localizing where one is looking at. An eye tracker is a device for measuring eye positions and eye movement. Eye tracking can be used to analyze visual attention, cognitive load and affective states. In gaze-controlled interfaces, eye gaze is used to directly operate a display
- **ASR/DVI**: Automatic speech recognition (ASR) and direct voice input (DVI) interpret spoken voice and use it as input to the interactive system. While recent advancements in LLMs enable electronic systems to interpret any spoken command, older systems and a few safety critical systems like aircraft cockpits work on recognizing a finite set of keywords as voice commands
- **Gesture recognition systems**: A gesture is a form of nonverbal communication or nonvocal communication in which visible bodily actions communicate particular messages, either in place of or in conjunction with speech. Gestures include movement of the hands, face or other parts of the body. A gesture recognition system deals with the automatic recognition of gestures and uses them as an input for interactive systems
- **Multimodal fusion and fission**: For interactive systems, multimodal fusion refers to combining and using multiple modalities of input (voice, eye gaze, head movement, text) channels into a single stream of input. Multimodal fission refers to providing output through multiple channels like changes in display, spoken output and haptic feedback Multimodal fusion and fission investigate algorithms to combine or produce different modalities of interaction to improve user experience compared to the unimodal system
- **Extended reality**: Extended reality (XR) refers to environments that combine the real and the virtual, through the use of computer technology and wearable devices. XR technologies consist of virtual, augmented and mixed reality (respectively, VR, AR and MR)

4.7 SUMMARY

In summary, this chapter provides a brief overview of different modalities of interaction. We focussed on modalities for which commercial-off-the-shelf

(CoTS) devices are available and did not discuss more emerging technologies like interfaces transmitting olfactory sensation. It may be noted that many new modalities of interaction like eye tracking, gesture recognition, and augmented or mixed reality interfaces depend on computer vision technology for real-life object detection, be it be pupil location for eye tracking, hand movement for gesture recognition or any random object in the case of AR/MR-based interaction. The next chapter will introduce basic theories of computer vision with a case study on developing an MR interface.

Chapter 5

Computer vision

5.1 INTRODUCTION

From its inception, one of the main aims of AI was to mimic the human information processing system. In the mid-1970s to early 1990s, the logical reasoning process was attempted to be simulated through discrete mathematics propositions and rules (e.g., A OR B AND C). However, there are several human information processing systems that work faster than can be explained through explicit symbol processing using logical operators. Pattern recognition, including face recognition, and natural language processing are a few such examples. Although the concept of neural networks emerged in the 1950s, it was not until the emergence of deep learning in the first decade of the 21st century that the power of AI was appreciated like never before through Nobel Prizes and the Turing Prize. One of the important applications of AI is the advancement in computer vision and enabling machines to see and understand the world as we humans do.

Just like the symbolic AI paradigm, the field of computer vision started with manually selecting facts and rules in the form of features and image processing algorithms. Later on, with the advent of image sensing technology, manual selection of features and image processing algorithms did not remain scalable. If we look back at the human visual pathways, discussed in Chapter 1, it may be noted that in our optical cortex, we indeed process individual visual features in the V1 and V2 regions, but we do it automatically, not consciously. A similar approach is taken in modern computer vision algorithms that automate the feature selection and image interpretation processes. In this chapter, we will investigate the basic computer vision pipeline, its modification from manual feature selection to deep learning-based systems and applications on developing intelligent interaction techniques discussed in the previous chapter.

DOI: 10.1201/9781003646716-5

5.2 CONVOLUTIONAL NEURAL NETWORKS (CNNS)

In the previous chapter, we discussed different modalities of interaction, including gesture recognition and eye gaze estimation. Whenever we require detecting objects of interest like forearms and fingers or the eye region from facial images, as of 2025, convolutional neural networks (CNNs) have been used as the de facto standard for object detection. In Chapter 2, we discussed the basic pipeline of visual processing. How can machines perceive like us? As we already discussed in the introduction section, humans are often better at pattern (including face) recognition than contemporary machines. Before the deep learning era (prior to 2010), computer vision worked by identifying feature detection algorithms, improving their performance and combining multiple feature detection algorithms to generate the power of vision in machines (Nixon & Aguado 2002). For example, the following three-layer feedforward neural network was used to predict eye gaze fixation duration from colour histogram and shape context coefficients for basic shape and icon-based stimuli on a white background. However, with the advent of cameras and communication technology, recording and transmitting images and videos became more ubiquitous, leading to automating the process of feature selection to develop computer vision systems. Researchers took inspiration from the biological process of vision of selecting relevant features from a limited receptive field.

Before going into details of CNNs, we should first understand the convolution operation. Mathematically, the convolution of two functions is defined as the integral of the product of the two functions after one is reflected about the y-axis and shifted. In image processing, we will discuss 2D discrete convolution, which can be imagined as the process of wiping a board with a duster or wiper. The image is represented by a 2D matrix of values indicating colour values at each position. A smaller matrix, known as a mask or kernel, is multiplied and slid through the image left to right and top to bottom. Mathematically, the 2D convolution of two input matrices is defined as:

$$C(i,j) = \sum_{m=0}^{(Ma-1)} \sum_{n=0}^{(Na-1)} A(m,n) * B(i = m, j - n)$$

where $0 \leq i \leq M_a + M_b - 1$ and $0 \leq j \leq N_a + N_b - 1$, where A and B are matrices of dimensions M_a, M_b and N_a, Nb, respectively. In image processing, convolution is often used for feature detection like edge detection. Generally, in a CNN, a series of repetitive neural networks try to learn appropriate kernels or masks for automatically detecting image features, followed by a fully connected feedforward network to do classification (Figure 5.1). The convolution layers reduce the number of branches and hence model parameters compared to a fully connected network taking the whole image as input. For example, if an image of 10 MP resolution is

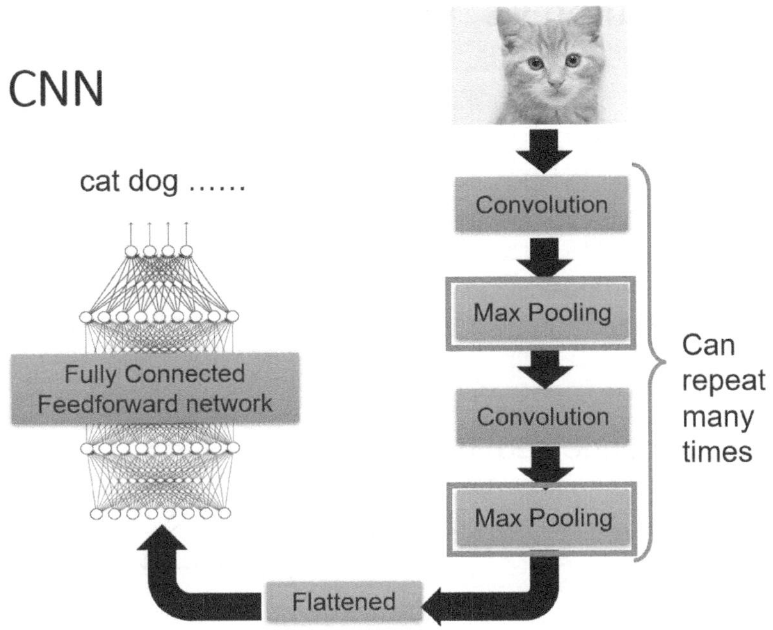

Figure 5.1 Structure of a representative CNN.

directly fed to a neural network as described in Chapter 3, we will have model parameters in multiples of 10 million, increasing the training time and chances of overfitting. In a CNN, we take multiple kernels over the images like the receptive fields of the visual cortex and reduce the total number of parameters (Figure 5.2) (Zhang et al. 2023).

While there are many variations in CNN architectures, the basic architecture consists of a:

Convolution layer: A CNN will have multiple convolution and pooling layers. Each convolution layer learns a mask or kernel during the training process and automatically identifies relevant features during the training process. It works by iteratively downsampling an image and localizing objects of interest. Figure 5.3 shows the status of an image as it passes through multiple convolutional layers

Pooling layer: The pooling layer downsamples (reduces spatial dimension) the intermediate image by dividing it into multiple grids and represents each grid by a representative value. The representative value can be the maximum value among the grid of values (max pooling) or the average value (average pooling). The pooling function mainly tells where a specific feature is present in a particular image region or not instead of accurately localizing it

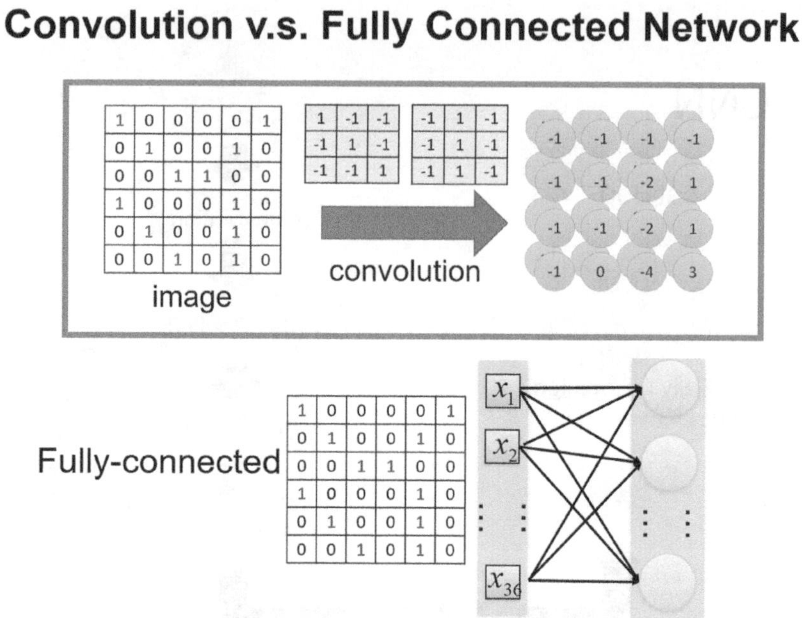

Figure 5.2 Model parameter reduction in a CNN.

Flattening and classification layer: After downsampling, the low-dimensional image is represented as a single array and fed to a feedforward artificial neural network, as discussed in Chapter 3, which undertakes the final classification of content

While there are many different types of CNNs available in the literature with new models emerging frequently, as of 2025, the most popular one is the Yolo series of models. The original Yolo (You Look Only Once) model worked by dividing an image into a set of grids and looking for objects of interest in each grid. The accuracy of object detection models is evaluated using similar criteria as classification systems (false positive, false negative), as discussed in Chapter 3. We also use another parameter known as intersection over union (IoU), which calculates the amount of overlap between two bounding boxes: a predicted bounding box and a ground truth bounding box. The ground truth bounding box is part of the input or training sample to an object detection model. While testing, the model outputs a bounding box to encapsulate an object of interest. Ideally, it should match the ground truth bounding box; a higher value of IoU indicates a better match than a lower value (Figure 5.4).

Chapter 10, Section 10.3 demonstrates an example of a CNN-based eye gaze estimation system. However, we shall move on to understand another interesting computer vision model known as the vision transformer.

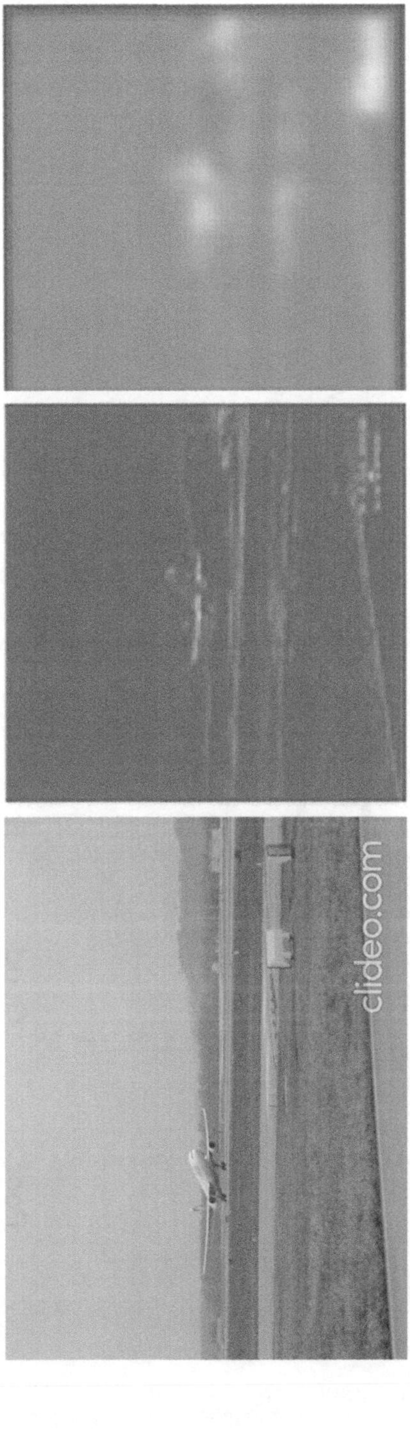

(a) Input Video

(b) 28th channel of the intermediate activation of the 4th layer

(c) 28th channel of the activation of the 9th layer

Figure 5.3 Visualization of intermediate layers of a CNN.

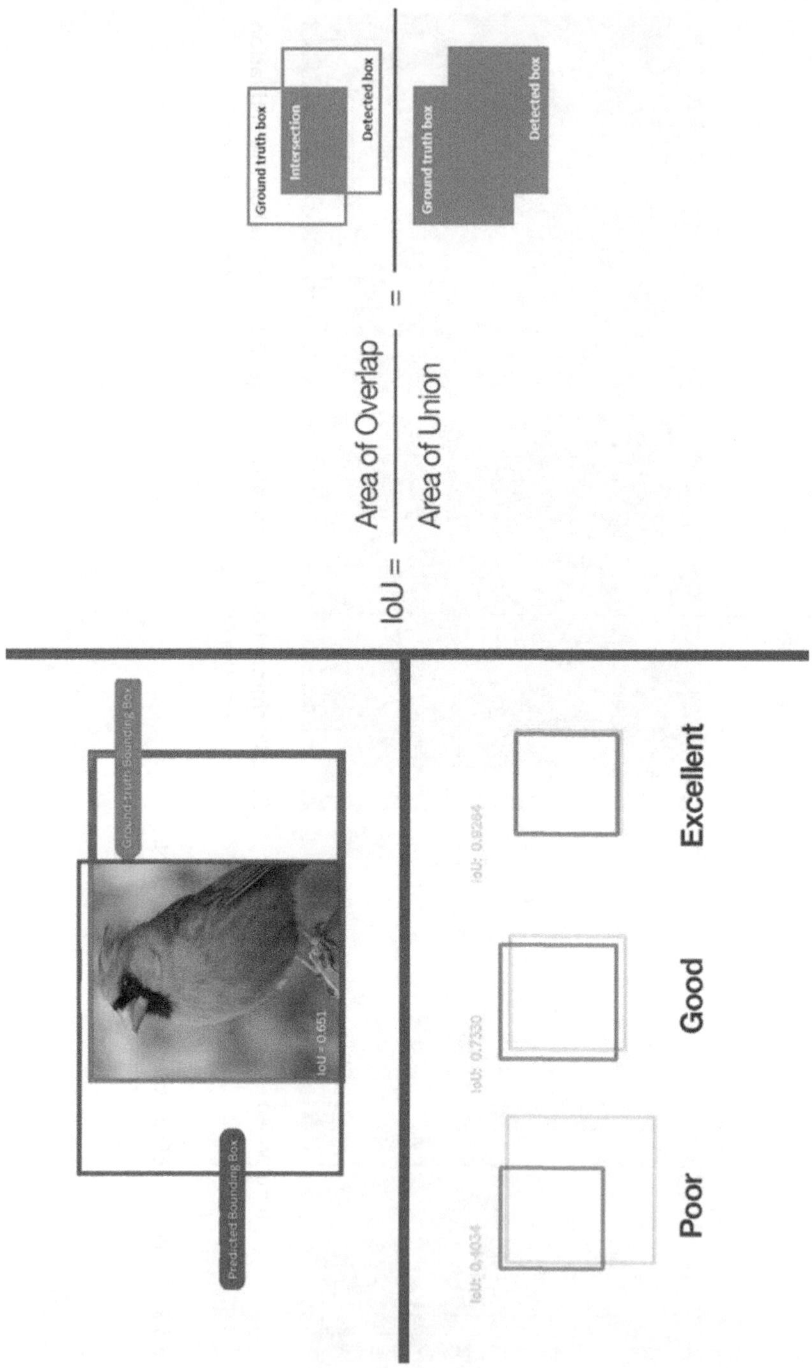

Figure 5.4 Explanation of the IoU metric.

5.3 VISION TRANSFORMERS

Vision transformers (Dosovitskiy et al. 2021) are an alternative approach to object detection than CNNs. The transformer model was proposed in the context of natural language processing in 2017 (Vaswani et al. 2017) and is discussed in detail in Chapter 6. Vision transformers (ViTs) work by the following steps:

1. Divide an image into multiple patches
2. Encode each patch into vector representation and separately encode the positional information of each patch within the image
3. Feed the encoded vectors of patches and positional information into a transformer encoder (Devlin 2018). The transformer architecture is discussed in the context of natural language processing in Chapter 6
4. Run a multilayer perceptron based on the output of the transformer encoder to classify

In CNN, the positional information of different segments of an image is retained through the process of the movement of the convolution mask over the image. However, in a ViT, the positional information is used at the starting point during positional embedding and during fine-tuning for downstreaming ''a model for a specific domain. As a result, the model is forced to 'learn' spatial relationships among different image patches. A ViT is less data efficient than a CNN as it requires a bigger size of training dataset, but once it is trained, the ViT is often found to be more transferable and robust for object detection or image classification tasks across multiple domains. Table 5.1 presents a comparative analysis between a CNN and ViT.

The original ViT architecture is modified for different applications, such as reducing the size of a training dataset, robust feature learning and synthetic data generation. The masked encoder runs two ViTs, with the second one decoding or regenerate the input image. The encoder block uses a subset of patches from an original image while the decoder learns to reconstruct the whole image from the partial input image. The masked autoencoder (He 2022) can reduce the size requirements of a training dataset and can also be used for inpainting applications. Like the masked autoencoder, Meta AI research released a model DINOv2 (Caron et al. 2021), which learns robust visual features through a curated and uncurated dataset constructed from the ImageNet dataset and web crawling. The DINOv2 model can be used with other computer vision or object detection models to improve robustness of the model. The ViT-VQGAN model learns a list of symbols from an image and can convert a list of symbols to an image. The list of symbols can be used to generate synthetic data as explained in the next section and

Table 5.1 Comparison between CNNs and vision transformers

Feature	*CNN*	*Vision transformer*
Architecture	• Relies on a hierarchical structure of convolutional layers, pooling layers and fully connected layers • Uses convolutional filters to extract local spatial features (e.g., edges, textures) from image regions	• Divides an image into fixed-size patches, flatten these patches into sequences and process them using transformer layers • Consists of multihead self-attention mechanisms and feedforward networks
Feature extraction	• Extracted through localized receptive fields, where filters slide over the image to capture spatial patterns	• Treats image patches as tokens, similar to words in natural language processing, and uses self-attention to capture global relationships between all patches • Allows the model to learn dependencies across the entire image without assuming local correlations
Locality bias	• Designed with a strong inductive bias toward local spatial relationships, leveraging the assumption that nearby pixels are more correlated than distant ones	• Lack a built-in locality bias and instead relies on attention mechanisms to model long-range dependencies, making them suitable for tasks where global context is critical
Processing mechanism	• Uses convolution operations to apply shared filters across the image, reducing parameters and enforcing translation invariance • Use of shared weights in convolutional kernels results in fewer parameters compared to fully connected layers, making CNNs scalable for large images	• Employs self-attention to weigh the importance of each patch relative to all others, enabling the model to capture global dependencies • Requires more parameters and computational resources due to the dense attention mechanism

chapter. As pointed out at the beginning of this section, transformers were initially proposed for natural language processing, and one can combine image and text encodings using ViT and standard transformers for textual data. For example, an image and its caption or description can be coded together using transformers, and then both image and text embeddings can be used for conversational agents, information retrieval and even for instructing robotic agents (Zhang et al. 2023). The system is known as the vision language model, and we will describe a use case of navigating a rover using both image and text prompts in Chapter 9.

5.4 SYNTHETIC DATA FOR MR APPLICATIONS

The previous two sections described computer vision models for detecting or classifying objects and subsequently using them for developing novel interaction techniques like eye gaze estimation or tracking objects for mixed/augmented reality systems. However, one disadvantage of machine learning-based computer vision models over traditional ones is the requirement of a large quantity of annotated data. The name mixed or augmented or extended reality systems often pushes people towards unity or unreal engine-based 3D content development, overlooking the fact that the success of any interactive system depends on user acceptance of the accuracy and latency of the system, and for MR/XR systems, the accuracy and latency depend on underlying computer vision system, not that much on 3D content. From 2010 onwards, most practitioners use deep learning-based computer vision systems requiring a large amount of training data for novel applications where pretrained models are not available. For defence and many industrial applications, it is often not possible to capture images to create a training dataset. Sometimes even the deployment scenario is unknown or not yet physically deployed. In recent times, synthetic data has turned popular not only for developing human-like content but also for training computer vision systems for unknown scenarios. Researchers already investigated different ways to generate synthetic data as follows:

- **Using 3D gaming engine**: Researchers explored using 3D gaming engines to generate synthetic data. There is reported work on generating traffic crossing situations for developing autonomous heavy vehicles and airport situations for automating taxiing of aircraft. Such situations are difficult to generate in reality for example, photographing another vehicle or aircraft on the runway (Gaikwad et al. 2023), but should be included in the training dataset for safe and correct obstacle detection and avoidance strategy. However, manually generating all possible scenarios in a gaming engine is resource intensive and often not scalable
- **GAN-based approach**: Generative adversarial networks (GANs) were a pioneering approach to synthetic data generation. Initial GAN models used two competing backpropagation neural networks generating and discriminating data from a training dataset. Image translation models based on GAN often consist of multiple GAN models and require large amounts of images, for some models paired images to start the synthetic image generation process. We will not discuss details of GAN here, but interested readers may check Sahu et al.

(2024) for details on the comparison of GAN models for synthetic data generation

- **Diffusion-based approach**: As of 2025, diffusion models are considered state-of-the-art for synthetic data generation. Diffusion models generate synthetic data by a forward pass of adding Gaussian noise to a dataset and a reverse denoising step. The process involves training a neural network to sequentially denoise images blurred with Gaussian noise; the denoising process generates new images with the original distribution of real images

While an exhaustive discussion on generative AI techniques is out of scope of this book, the next section describes a case study of developing an MR-based intelligent user interface using concepts of generative AI and computer vision.

5.5 CASE STUDY OF GENERATIVE AI AND COMPUTER VISON

In this section, we will present a novel case study of generating synthetic data using a diffusion model and then briefly describe a mixed reality-based application that used a computer vision model trained with synthetic data but performed in real time with physical objects. We picked a case study in the context of Industry 4.0. The particular application aims to develop an MR-based application to help factory workers assemble and disassemble systems like automotive or aircraft parts. The challenge involves the non-availability of suitable annotated images in real factory floors or workshops. Initially, we proposed a diffusion model-based synthetic data generation pipeline for different machine tools and components with a wide variety of sizes and orientations, and next, we described an MR application to assemble components of a pneumatic cylinder. Details of the project can be found in Sinha et al. (2025).

The proposed method uses an image-to-image latent diffusion pipeline (Figure 5.5). The foregrounds of the segments of the tools are composed into relevant backgrounds and refined using RealVis XL. A low noise injection level – corresponding to the final ten denoising diffusion implicit model (DDIM) steps – preserves the original pose and structure. Material details are enhanced while geometric fidelity remains intact. The decoder predicts shadows, depth cues and interreflections.

5.5.1 Synthetic data generation pipeline

Foreground collection and mask generation: The selection of the foregrounds encompasses a range of tools that vary in size and function, including frequently utilized hand tools as well as smaller fastening and

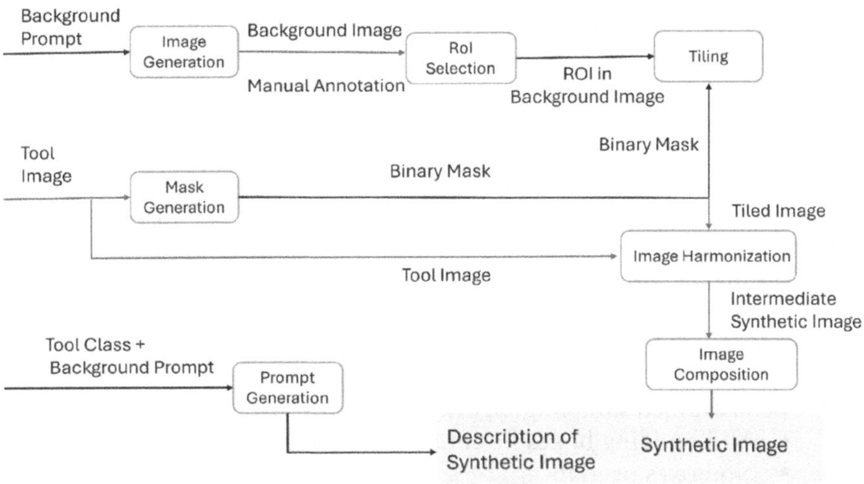

Figure 5.5 A synthetic image generation pipeline that combines tool foregrounds with text-guided background prompts to generate annotated, photorealistic tool datasets.

mechanical components. These tools were chosen for their widespread applicability in smart manufacturing systems, where precision, modularity and integration with robotic and cyber-physical systems are essential. The selected 16-tool dataset comprehensively supports such workflows by covering tasks like engine housing assembly and gearbox maintenance – two representative processes that collectively utilize nearly all tools in the dataset. A dataset of tool instances was constructed using an automated image collection pipeline, tailored for each tool category. The process involved issuing precision-targeted search queries across public image search engines and retaining the top N results per query.

- Heuristic-based filters were applied to eliminate watermarked, low-quality or non-photographic images
- The remaining RGB images were processed using an image segmentation model to generate pixel-accurate foreground masks

Image generation: A total of 30 expert-crafted prompts were developed to simulate a wide range of spatial and visual contexts. These prompts were designed by manipulating semantic attributes such as time of day, weather conditions and indoor versus outdoor scenes, including industrial workbenches, outdoor repair zones and factory interiors.

- SDXL-generated backgrounds with diverse hue, saturation and value (HSV) characteristics that influenced shadows, reflections and material cues

- The model's capability to generate high-resolution and diverse visual scenes contributed to improved variability in the dataset and reduced overfitting

Role selection and tiling: Tool foregrounds after masking are composited onto background images using the region of interest defined by annotators. These constraints are defined as polygonal regions drawn manually on each background image, specifying the area within which tool instances can be placed. This region is chosen to fit one or more square bounding boxes for tool placement. For multi-instance compositions, five tools are placed per image. Two tools are selected from the 'big' category and three from the 'small' category. All tools are sampled from distinct classes.

- A constrained sampling algorithm ensures that:
  - All bounding boxes lie entirely within the annotated polygon
  - No boxes overlap
  - Minimum spacing between tools is maintained using adaptive heuristics
- A bottom-biased placement strategy is applied to mimic realistic tool arrangements

Image harmonization: Luminance consistency between the foreground and background is essential for realistic composition. The brightness of the foreground is adjusted using HSV-based normalization by scaling its value channel to match the average brightness of the background. Hue and saturation remain unchanged, and the scaling factor is clipped to maintain a natural appearance. This step improves perceptual coherence and leads to more seamless blending during diffusion-based refinement.

Image composition: The intermediate synthetic image (harmonized image) is sent as an input to a diffusion model with a low strength parameter (0.2) and a guidance scale of 10. The low strength preserves the tool's geometry and texture, while the guided diffusion subtly resamples high-frequency details to blend the foreground naturally into the scene without introducing structural distortions.

Multi-Modal Annotations: Each refined synthetic image is paired with auxiliary annotations that enable broad applicability across visual recognition tasks. Bounding boxes facilitate object detection, segmentation masks provide detailed spatial structure and text prompts offer semantic grounding for vision-language models. These layers enhance the dataset's compatibility with modern detection, segmentation and multimodal frameworks:

1. Bounding boxes: Derived from the foreground mask and stored in YOLO format, ensuring compatibility with single-shot detectors and minimal background leakage

2. Segmentation masks: Provided at full resolution as lossless PNG files to maintain pixel-level fidelity, supporting high-accuracy instance and panoptic segmentation
3. Text prompts: Generated via a three-stage process: (1) sampling from ~2000 scene descriptions, (2) inserting the tool name as a fixed token and (3) expanding with Flan-T5 XL for semantic and stylistic variation, enabling reproducible text–image alignment

5.5.2 MR application using synthetic data

An MR application was developed to provide instructions to the user illustrated in Figure 5.6. The manual assembly of pneumatic cylinders had four components – bottom cover, cylinder, piston and top cover – as illustrated in Figure 5.6. During the task execution, the assembly components and the marker were placed on the table. The process entails that the user:

1. Fix the bottom cover
2. Place the cylinder on the bottom cover
3. Insert the piston inside the cylinder
4. Fix the top cover

The process begins with the user donning the Hololens2, a mixed reality device, and standing in front of the assembly area. Multiple buttons are provided to the user. First, the user takes a picture of the workspace. The image is then transmitted to the computer for inference, where the model generates the bounding box positions for each class. After getting the bounding box position, homography is used to achieve precise spatial transformation from pixel space to <x,y,z> coordinates so that the holograms appear appropriately for different relative positions of the operator donning the mixed reality headset. Homography is a mapping between two planar projections of an image.

Once the <x,y,z> coordinates are obtained, holograms corresponding to the given parts appear next to the parts. Once all the virtual objects are rendered onto the real objects, the user could click on the Next and Prev buttons to follow the assembly instructions. Initially, all the virtual components are rendered in blue colour. When the user clicked the Next button, the colour of the components that needed to be assembled next would change to green. The view of the entire assembled component is also rendered onto the workspace in green to the left. A yellow-coloured arrow was rendered on the workspace, where the tail end indicated which component to pick. The task was terminated when all components were assembled. A user study involving 11 end users found that users could finish the assembly of one pneumatic cylinder within 55 seconds, with an average TLX score of less than 25 on a scale of 100 and a Cronbach alpha score of 0.8 indicating convergence of the learning experience.

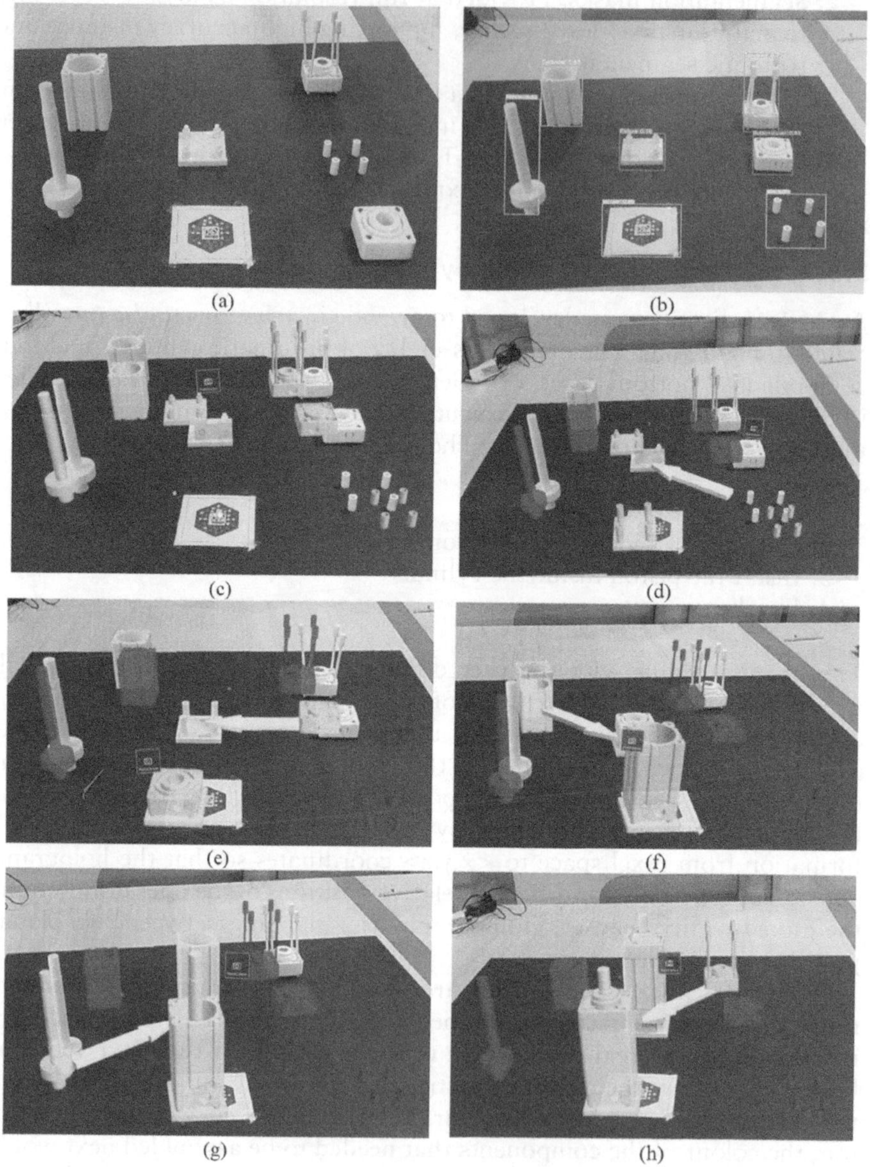

Figure 5.6 Step-by-step guidance information is given to the user during the assembly task. The 3D-printed components are in white. Virtual holograms are rendered in blue, while the hologram of the object that needs to be picked next is rendered in green. Yellow directional arrows are also provided.

5.6 QUICK FACTS

- **Classical computer vision**: It is a way for computers to understand images using basic rules and patterns, like finding edges or shapes, without using learning like humans do. Think of it as teaching a computer to 'see' using step-by-step instructions
- **CNN (convolutional neural network)**: It is a type of computer model that learns to recognize patterns in images, like faces or objects, by looking at many examples. It works a bit like how our brain processes visual information. It is called 'convolutional' because it uses a mathematical technique called convolution to scan and detect features in different parts of the image
- **Bounding box**: This is a rectangle drawn around an object in a picture to show where it is. For example, it can highlight a person, a car or a dog in an image. Bounding box is a common output of object-detection models
- **Segmentation**: This means breaking an image into parts so the computer can understand what each part is, like separating a person from the background. It is like colouring in each object with a different crayon to tell them apart. Segmentation can show the exact contour of an object, while the bounding box only draws a box surrounding an object
- **Vision transformer**: This is a newer model that helps computers understand images by focusing on different parts of the image, kind of like how humans pay attention to important details. It's like scanning a picture piece by piece to get the full idea. It represents an input image as a series of image patches, like the series of word embeddings used when using transformers to text and directly predicts class labels for the image

5.7 SUMMARY

In summary, this chapter introduces basic concepts of computer vision and its utility in intelligent interaction design. The field of computer vision is fast progressing with new models or architecture being published almost every month. We did not dive into the details of any model or specific architecture; rather, we presented the input, output and context for using different models. We expect the reader to understand why deep learning-based models are prevalent now and the different types of output from models like bounding boxes, segments or semantic details. In the upcoming chapter, we shall learn to process textual information and the evaluation metrics for measuring the accuracy of retrieved information, which will also be applicable for computer vision systems.

Chapter 6

Textual information processing

6.1 INTRODUCTION

The previous chapter discussed processing image or video data and interpreting them to develop new input (gaze-controlled interface, gesture recognition system) or output (augmented or mixed reality) interactive systems. This chapter will discuss processing textual data and interpreting it for developing intelligent interactive systems. In the late 1960s and early 1970s, expert systems were one of the most popular AI systems predicting chemical composition (Dendral; Lindsay et al. 1993) or even prescription medicine (Mycin; Shortliffe & Buchanan 1975) from textual responses from users. With the emergence of the World Wide Web in the 1990s, text and hyperlink processing gained momentum not only in terms of giving access to information but also creating information for the audience. This creation part took a new turn with the emergence of social networking and microblogging sites. The development of machine learning models often worked vis-à-vis with image and textual data analysis – for example, LSTM (long short-term memory) and transformer models were originally developed for textual content analysis but later used for computer vision systems as well in the form of vision transformers. Presently, a major research area of intelligent interaction design is centred on using large language models (LLMs) and combining them with computer vision models to understand users' instructions and context. This chapter will start from basic content analysis using stemming, stop words, parts of speech tagging and the concept of a vector space model. Next, we will discuss expert systems in the context of online recommendation systems and chatbots. Finally, we will seek to understand the basic concept and structure of LLMs.

6.2 TEXT MATCHING

The first section of this chapter will discuss how we match similarities between textual information, and, in the process, we will also discuss the preprocessing of textual data before advanced inferencing. If we type a

DOI: 10.1201/9781003646716-6

natural language expression in a search engine, conversational agent or any other textual inferencing system, it traditionally uses the following processes (although in Section 6.4 we will discuss more advanced text processing techniques):

- **Stop word removal**: A set of common words is removed from the text. Stop word lists for different languages can be pre-stored in the database. Common examples of stop words are prepositions, articles and pronouns. We can include a stop word in subsequent searches by putting double quotes around it for most web search engines
- **Stemming or lemmatization**: Stemming or lemmatization removes suffixes from words so that different version of same word is similarly processed. For example, compute, computing and computed are similarly processed. Stemming can be implemented based on specific rules for a specific language (for example, Porter's algorithm for English) or it can use part-of-speech (PoS) tagging. The lemmatization process is more accurate than rule-based stemming, but PoS tagging is often resource intensive for different languages
- **Term frequency**: After stop word removal and stemming, we can count the occurrences of each term in a document and generate a frequency list for all terms. The raw term frequency list will show bigger values for longer documents and commonly used terms for a specific corpus (for example, the word 'computer' in ACM DL). Hence, the raw count of terms is normalized with respect to document length and uniqueness of the term. The uniqueness is measured through inverse document frequency (IDF), which is the reciprocal of the number of documents in which a specific term appeared
- **Vector space representation**: A corpus of documents is represented as a vector space with one axis listing all documents in the corpus and another axis representing all terms in the corpus. Within this space defined by documents and terms, each document and a query or search string is represented as a vector with values indicated by normalized term frequencies. Standard vector similarity metrics like dot product or cosine similarity are used to find similarities between documents or the similarity of a query/search string with existing documents (Salton et al. 1975)
- **Evaluation**: How can we evaluate the accuracy of an information retrieval system? Let us consider that we have a corpus with many documents; some of them are relevant for a specific query. Ideally, a search engine should retrieve all relevant documents. However, in reality, the search engine retrieves a set of documents, and only a subset of these retrieved documents is relevant (Figure 6.1). We use similar metrics as we used for measuring the accuracy of classification systems (refer to Chapter 3, Section 3.6) – precision and recall

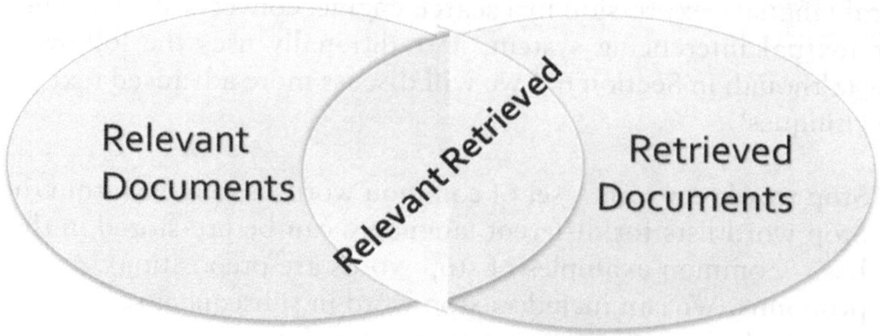

$$F_1 = 2 \cdot \frac{1}{\frac{1}{\text{recall}} + \frac{1}{\text{precision}}} = 2 \cdot \frac{\text{precision} \cdot \text{recall}}{\text{precision} + \text{recall}}$$

Figure 6.1 Precision and recall explanation.

- Precision = Retrieved relevant documents/Retrieved documents Recall = Retrieved relevant documents/Relevant documents
- Recall = Retrieved relevant documents/Relevant documents

Precision and recall often have a trade-off: if we optimize a system for high precision, the recall will be less and vice versa. For example, we will have high precision with less stemming but better recall with stemming. Many systems define a function like F1 score, combining both precision and recall to try to optimize the combined function.

6.3 EXPERT SYSTEMS

The previous section discussed matching a query with documents from a corpus. However, many times we would like to undertake a conversation beyond a single search string. When we book a ticket or complaint, or debug a software problem with an automated agent, we make use of such a conversational agent. In Chapter 3, we briefly discussed about logical reasoning system in the context of a symbolic versus connectionist approach to AI. In this section, we shall briefly explain expert systems, which were popular in the past century and are still used in specific use cases like legal case processing. Expert systems were originally developed by hand and most commonly in Lisp. It was discovered that many problems were being solved by chaining through rules (if–then statements) that would operate on a collection of facts and partial conclusions. These rule-based systems led to the first AI tools or shells. A basic rule-based expert system consists of:

- **Assertion**: A statement about a fact
- **Rule-based system**: A collection of if–then rules

- **Working memory**: A collection of assertions that are matched with if-patterns
- **Deduction system**: Then-patterns specify assertions to be placed in working memory
- **Reaction system**: Then-patterns specify actions

The expert system takes input from users, and the input is added to working memory. Next, the rules in the rule-based system are fired (Figure 6.2). The sequence of rule firing is based on reasoning system. Two popular approaches are forward chaining (reaching from assertions to a conclusion) or backward chaining (going back from the desired goal to the required assertions). If one has not yet gathered any facts, and is only interested in whether one of many possible conclusions is true, backward chaining should be used. For example, if we are only interested in whether an animal is a carnivore or not, we shall only look at antecedents of the carnivore rule to focus fact-finding. If one has all the facts that one will ever get and is interested in everything that one can conclude from those facts, forward chaining should be used. For example, if we catch a fleeting glimpse of an animal, gathered a set of facts and no more facts possible because the animal is gone, we can use forward chaining to identify the animal.

```
(rule (if has-sauce is yes and
          sauce is sweet)
      (then best-sweetness is sweet with certainty 90 and
            best-sweetness is medium with certainty 40))

(rule (if preferred-sweetness is dry)
      (then best-sweetness is dry with certainty 40))

(rule (if preferred-sweetness is medium)
      (then best-sweetness is medium with certainty 40))

(rule (if preferred-sweetness is sweet)
      (then best-sweetness is sweet with certainty 40))

(rule (if best-sweetness is sweet and
          preferred-sweetness is dry)
      (then best-sweetness is medium))

(rule (if best-sweetness is dry and
          preferred-sweetness is sweet)
      (then best-sweetness is medium))
```

Figure 6.2 Examples of rules in a rule-based system.

6.4 LARGE LANGUAGE MODELS

The term-frequency calculation discussed in Section 6.1 considers number of occurrences of a term and its uniqueness through the inverse document frequency, however, it misses the context of occurrence of a term. For example, if we consider these sentences – *I went for a walk at the riverbank* and *I was going to the bank for cash withdrawals* – the different meanings of the word bank are quite obvious from the rest of the context. On the other hand, *I went for a walk at the riverbank* and *I was running at the riverbank* bear similar meanings, but the verbs 'walk' and 'run' are syntactically different. It is not only for synonymous or homonymous words, but in general context offers a lot of information while we interpret natural language (Figure 6.3). In the case of expert systems, although they were popular in the last century, the handcrafted rules were not scalable, just similar to classical computer vision discussed in previous chapter. It will be ideal if we can automatically capture context and reasoning logic from the text itself. We have one advantage in recent times in terms of availability of large corpora thanks to the widespread use of the World Wide Web and ease of digital content development.

The SMART (System for the Mechanical Analysis and Retrieval of Text) Information Retrieval System developed at Cornell University in the 1960s first used the vector space model (Salton et al. 1975) and later it was extensively used in information retrieval, indexing and so on. The task of information retrieval and indexing is gradually subsumed by the bigger field of natural language processing (NLP). When we learn a new language, we start with the alphabet, then vocabulary and then grammar. Earlier attempts in language modelling took a similar approach following Chomski's phenomenal work in 1950s to use formal grammars to model language. Researchers used similar concepts of formal grammar to develop programming language, compiler and even user models (task action grammar).

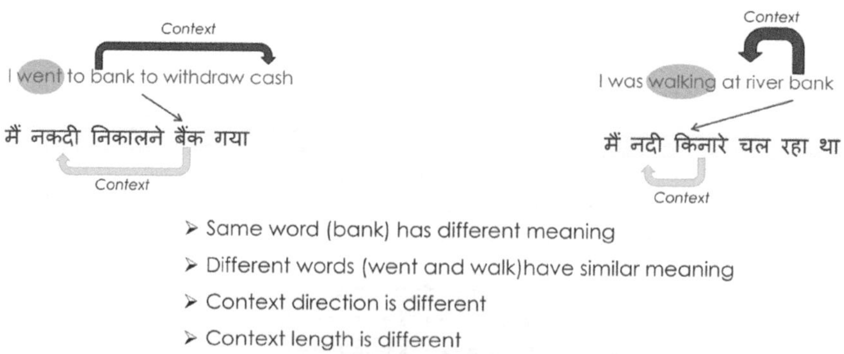

Figure 6.3 Explaining context during machine translation.

The earlier attempt of language models treated the words or terms as in the form of various n-gram models, where each word is considered as a unit (gram), and statistical models were used to elicit relationships among words. However, as we saw multiple times in the context of the symbolic and connectionist approaches of AI and classical versus CNN-based computer vision, researchers explored use of neural networks to automatically model language. In 1986, Rumelhart pioneered a concept of embedding next-of-kin relationships in a backpropagation neural network, where the network through its weight adjustment learns the next-of-kin rules. A specific type of neural network, called a recurrent neural network (RNN), found application to model the context of words. Unlike feedforward neural networks discussed in Chapter 3, recurrent neural networks use feedback connection where the output of a neuron at one time step is fed back as input to the network at the next time step. This enables RNNs to capture temporal dependencies and patterns within sequences. The Word2Vec model, published by Google researchers in 2013, used a two-layer neural network to make vector representations of words from a large corpus. Unlike the vector space model based on term frequency, the Word2Vec model (Mikolov 2012) was able to model semantics (or meaning) of words from context (surrounding words). An extension of the Word2Vec model, the Doc2Vec model, could represent each document or chunk of text as a vector like the original vector space model discussed in Section 6.1. The Word2Vec model captured context from a fixed-sized bag of words (unordered set of words). Later models like ELMo and BERT (Devlin et al. 2018) used longer lengths of input sequences in both directions (forward and backward) of a word to capture context using multiple neural networks (LSTM to be particular), which could disambiguate homonymous words and lead to the development of modern LLMs.

Another seminal paper by Google researchers in 2017, 'Attention is all you need', introduced the concept of transformers that is still now the de facto standard in LLMs, natural language generation and interpretation (Vaswani et al. 2017). The original paper was written in the context of machine translation, but authors already speculated further extension of the concept for non-textual and multimodal data. The transformer models, including the popular generative pre-trained transformer (GPT) models (OpenAI 2023) are based on a multiheaded (parallel) attention mechanism. The attention mechanism, in a nutshell, can be assumed to be a weighted average scheme. It was popularized in the context of NLP in 2014–2015 as a way of encoding context from a large corpus together, instead of one batch at a time like RNN-based models. The attention mechanism can be considered to have three steps:

1. A reader reading raw input and converting it into a distributed representation by creating a feature vector for each position of a word (minimum unit of prediction)

2. A working memory that stores all feature vectors and allows retrieving them in any order irrespective of initial order or sequence
3. A process that can selectively assign weight or attention to a specific feature vector

Say, for example, we are translating from a free word order Indian language to English. In most Indian languages, the relative positions of different parts of speech are not fixed, which is not true for English. The word-by-word translation may create grammatically wrong sentences, and writing handcrafted rules is also not scalable. The attention mechanism (Bahdanau et al. 2014) allows for assigning weights to different combinations of output sequences, and overtraining the right sequence will have more weights than others. The transformer models can undertake attention calculation in parallel for multiple combinations of input/output sequences leveraging modern GPU architecture. Once trained on a large-enough corpus, they get the ability to both interpret and generate human-link texts that can subsequently be used to develop recommender systems and conversational agents. The same system can also be trained on domain-specific corpus to make expert systems, as discussed in Section 6.2 without handcrafting rules (Figure 6.4). Readers may recall the discussion in Chapter 5 (Section 5.3) in the context of vision transformers, where individual image patches are encoded as tokens. We will discuss using a LLM to semi-autonomously control a rover in Chapter 9.

6.5 QUICK FACTS

- **Stemming:** It is the process of reducing inflected (or sometimes derived) words to their word stem, base or root form – generally a written word form. For example, the Porter algorithm reduces argue, argued, argues, arguing and argus to the stem argu

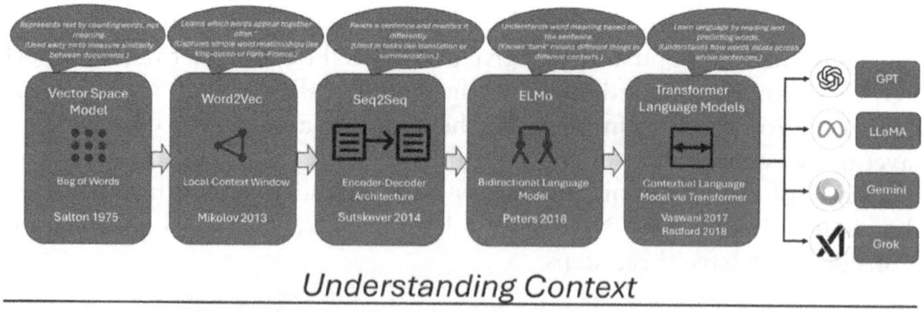

Figure 6.4 An organization of popular NLP systems based on increasing level of understanding context.

- **Vector space model:** Also known as a term vector model, it is an algebraic model for representing text documents (or more generally, items) as vectors such that the distance between vectors represents the relevance between the documents. It is used in information filtering, information retrieval, indexing and relevancy rankings
- **Precision** (also called positive predictive value) is the fraction of relevant instances among the retrieved instances, whereas **recall** (also known as sensitivity) is the fraction of relevant instances that were retrieved
- **Attention mechanism:** Inspired by the human capability to selectively perceive part of a stimulus, the attention mechanism can assign weight to one component in a sequence relative to other components in that sequence. Initially used for understanding the context of words in natural language processing, it is now applied in a wide variety of tasks including but not limited to machine translation, computer vision and speech recognition
- **GPT:** The generative pre-trained transformer is a large language model developed by OpenAI based on the transformer architecture. An off-shoot of GPT 4.0 is ChatGPT, which is fine-tuned as a conversational agent

6.6 SUMMARY

This chapter took us through basic techniques of textual content analysis and evaluation with pointers through contemporary research on LLMs in the context of intelligent user interaction design. Although the chapter is different than the previous chapter, readers are encouraged to find similarities between the process of image and text processing and evaluation as both processes work together in modern interactive systems. A plethora of resources is available on text processing for English and a few other popular languages, but developing corpora and corresponding analysis tools for non-English languages involves new challenges and opportunities.

Chapter 7

Information visualization

7.1 INTRODUCTION

In previous chapters, we learned how to analyze raw image data or text and elicit relevant machine and human-readable information. In this chapter, we will discuss how to visualize information for comprehension by humans (not machines!). As we studied in the first chapter, the human perception system has certain limitations as well as advantages. For example, we struggle to visualize data at higher dimensions, while we are often better at recognizing visual patterns than machines. Information visualization tries to exploit the human perceptual system to help decision-making. The field is vast, comprising product visualization, data analysis, theories of topography and cartography. The field is also rich in history, decorated with many important and lifesaving case studies like Dr John Snow's visualization of the cholera outbreak in London and subsequent analysis of the cause. The chapter will start with a brief historical perspective, followed by a description of basic visual primitives and a prospective visualization pipeline or architecture. As our focus will be on intelligent interaction design, we will describe an interesting user study comparing 2D and 3D visual primitives and developing a cognitive load dashboard. We will conclude the chapter with examples of tactile visualization.

7.2 INTERACTIVE DATA VISUALIZATION

Information visualization or data visualization explained facts and helped to make informed decisions from the beginning of the modern era. A few examples of famous visualizations were Edmond Halley's use of isolines to show contours of equal value on a coordinate grid, Dr John Snow's deduction that a cholera epidemic was caused by a bad water pump from the visualization of the location of deaths on a London map and Charles Minard's illustration of the disastrous result of Napoleon's failed Russian campaign of 1812. Visualization is explored for various applications like cartography and geovisualization, product visualization, information visualization and

DOI: 10.1201/9781003646716-7

high-dimensional data visualization. In this chapter, we will concentrate on interactive visualization, emphasizing examples of changing and analyzing visualization in real time.

Graphical objects are building blocks of visualizations, and identifying appropriate objects for a specific data type is crucial in developing user-driven visualizations. Basic graphs and graphical primitives can be amalgamated to form a complex visualization structure that is often difficult to interpret by users. Recognizing suitable graphs and graphical objects that form complex visualizations can help users to easily interpret information from those visualizations. In the context of visualization, a visual variable is an aspect of a graphical object that can visually differentiate it from other objects and can be controlled during the design process. Examples of visual variables can be size, shape, colour, opacity and position. While we render visual variables, a geometric primitive (or prim) is the simplest (i.e., 'atomic' or irreducible) geometric shape that the system can handle (draw, store). A common set of two-dimensional primitives includes lines, points and polygons, although some people prefer to consider triangles primitives because every polygon can be constructed from triangles. All other graphic elements are built up from these primitives. In three dimensions, triangles or polygons positioned in three-dimensional space can be used as primitives to model more complex 3D forms.

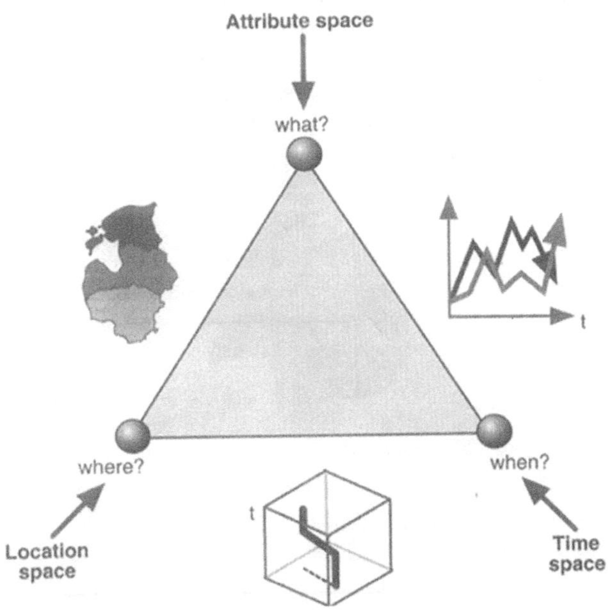

Figure 7.1 Three dimensions of geovisualization.

Interactive visualization can be defined through three dimensions: structure, navigation and description. Structure involves arranging the information into a hierarchy. Navigation refers to how the user moves through different levels of detail. Description is how the information is displayed or spoken, indicating how much information is conveyed. A similar construct is used for geovisualization – information is visualized through attribute space (what), time space (when) and location space (where; Figure 7.1). The user interaction changes one of these spaces – for example, when we tap on an icon on Google Maps, it shows more information about a location (attribute space), or when we double tap, we zoom in on a location (location space). In the following video, we used eye gaze tracking to navigate through a Google Map (https://youtu.be/NQKi5EiKKEc); fixation duration longer than 1 second at the central part of the map was used to zoom in, while conscious or exogenous eye blink was used to zoom out. Eye gaze fixation at the edge of the map was used to shift the map.

7.3 DATA VISUALIZATION PIPELINE

In this section, we propose a pipeline for interactive data visualization (Figure 7.2). The raw data are given as input to the start of the pipeline, and the data transformation stage then processes these data with analysis-centric methods like dimension reduction, subspace clustering and feature extraction to get the transformed data. The visual mapping process converts the analysis result from the data transformation stage or the original dataset into visual structures for rendering in the view transformation stage. The view transformation process handles what we ultimately see on the screen. It maps coordinates of points and lines that form the picture

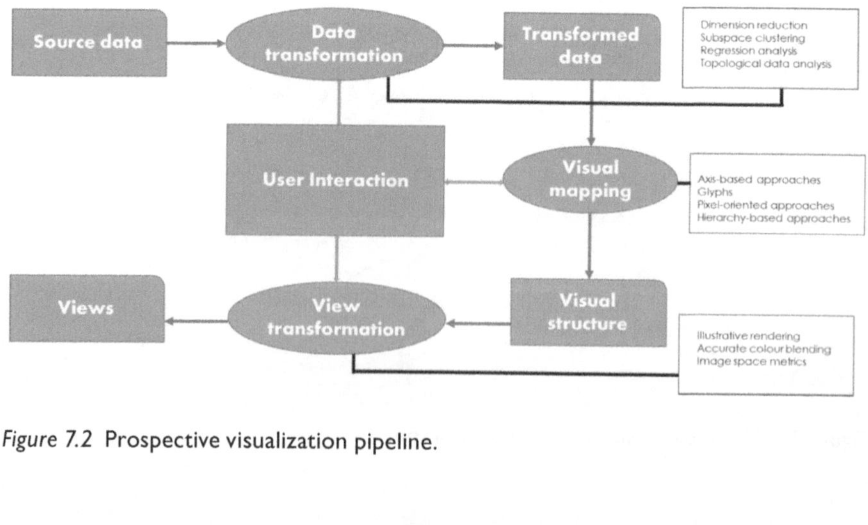

Figure 7.2 Prospective visualization pipeline.

into appropriate coordinates of the display device. Users can interact with three steps:

1. **Data transformation** – User interaction in this stage can serve several purposes:
   a. **Selecting appropriate techniques:** Choosing among various data transformation methods like PCA, t-SNE or clustering
   b. **Reducing features/dimensions:** Deciding which features to keep or ignore, and selecting how many dimensions to retain
   c. **Applying filters:** Selecting specific regions, groups or time intervals from the dataset to focus on for visualization

For example, Orange Data Mining is a GUI-based tool that supports drag-and-drop widgets for PCA, t-SNE, clustering and more. It also supports interactive selection of features to include or exclude. Scikit-learn is a Python library that supports APIs for PCA, t-SNE, UMAP and feature selection. Developers or users of a web interface can programmatically allow interactive feature selection.

2. **Visual mapping** – User interaction can include:
   a. **Choosing visual representation:** Selecting the type of visualization, such as scatter plots, bar charts or parallel coordinates. For example, using a line graph to show heart rate and bars to represent sleep patterns
   b. **Mapping data features:** Choosing how data variables are encoded visually, e.g., mapping frequency to size, category to colour or intensity to opacity. For example, selecting different categories using different colours and representing higher intensity values with higher opacity

3. **View transformation** – User interaction may include:
   a. **Zooming, panning, or rotating** the view – particularly useful for exploring 3D visualizations
   a. **Adjusting the camera view or angle** to focus on specific parts of the visualization
   b. **Clicking on a point** to inspect detailed underlying data or zooming into dense regions to explore local patterns

In summary, the objectives of information visualization can be summarized as:

- Make large datasets coherent – present huge amounts of information compactly
- Present information from various viewpoints
- Present information at several levels of detail, from overviews to fine structure

- Support visual comparisons
- Tell stories about the data

The next section will describe a user study of comparing visual primitives in 2D and 3D rendering media and using the information to develop an application to visualize cognitive load in real time.

7.4 USER STUDY TO COMPARE VISUAL PRIMITIVES

This study investigated and compared graphical primitives of 3D charts in virtual reality (VR) and 2D environments. We designed and conducted a user study with six types of 3D visualization techniques on both VR and 2D rendering media. Each 3D graph used in the experiment displayed numerical and nominal data using different combinations of visual variables. We considered a situation of displaying readings from multiple sensors in a built environment. We used the following six types of charts where the node represented the type of sensor (Figure 7.3), and a particular visual variable indicated numerical values of the sensor readings:

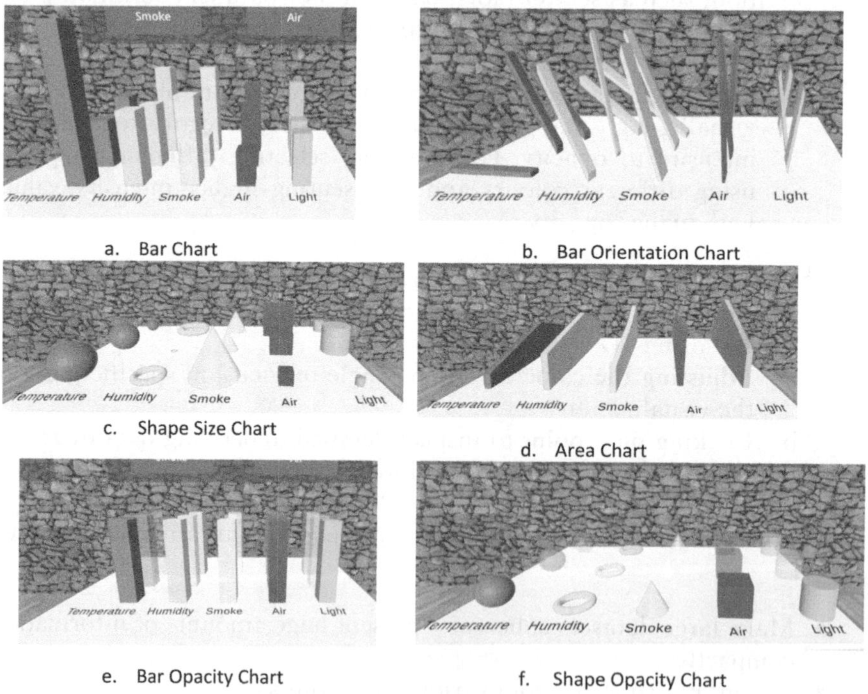

Figure 7.3 Types of 3D graphs used in the study.

- **Bar chart (BC)**: The nodes are represented by different colours, and the size of bars depicts a numerical value
- **Bar orientation chart (BOR)**: The nodes are represented by different colours but numerical values of sensors are defined by the orientation of bars. Bars are oriented or rotated along the x-axis to display values of sensors
- **Bar opacity (BO)**: The nodes are represented by the unique bar colours, and the opacity of bars is directly proportional to the numerical value of sensors. The darker the bar, the higher its value. The real value of the sensor is mapped to the opacity of the bar
- **Shape size chart (SS)**: The numerical values of the sensors are represented by the volume of the shape
- **Shape opacity chart (SO)**: The nodes are represented by a combination of shape and colour. Numerical values are defined by the opacity of the shape
- **Area chart (AC)**: Sensors are represented by planes in the chart and each sensor has a unique colour. The values of sensors are depicted by the peaks of planes. Each plane is scaled along the y-axis

We used HTC Vive Pro Eye with an inbuilt eye tracker and refresh rate of 90 Hz to collect gaze data and pupil diameter (accuracy 0.5° of visual angle at ideal conditions) on VR display. The Tobii-X3 120 eye tracker with a sampling rate of 120 Hz was used to collect eye-tracking data and pupil diameter (accuracy 0.4° of visual angle at ideal conditions) on 2D display. We also used the Emotiv Insight EEG tracker with five dry electrodes and sampling rate of 128 samples per second (SPS) to collect EEG data. My computer architecture consists of an Intel Core i5 processor and Nvidia 2070 graphics card. Data was collected from 20 participants (15 males and 5 females) with an average age of 27 years. Table 7.1 shows a summary of results. A bar chart with different columns for different values of data points was found to generate the most accurate response on VR display, while the area chart has the highest accuracy on 2D display. Bar-size charts induced the least cognitive load among users on both displays. VR display is found to be better than 2D display in terms of correct data interpretation.

Details of the study can be found in Arjun et al. (2021). The results on 3D graphical primitives were used to develop a VR cognitive load estimation dashboard using the Unity 3D game engine. The dashboard's main menu is divided into three modules: 2D, 3D and complex graphs. This kind of modular dashboard can leverage the advantages of both 2D and 3D graphs and help analyze gaze behaviour and cognitive load from a different perspective. Based on the previous study, 3D bar and area charts were used to investigate cognitive data, and all graphs used size primitives to represent numerical data, while colour was used to depict nominal data. Figure 7.4 shows the 2D and 3D interfaces of the dashboard. The 2D interface of the

Table 7.1 Similarities and differences between two displays

Parameter	*Similarities*	*Dissimilarities*
Percentage of correct answers	AC, BC and BO are best in terms of correct responses among six charts on both displays.	BC is the highest accuracy chart on VR display, while AC is the highest accuracy chart on 2D display.
Response time	On both displays BC has the lowest response time, while BOR has the highest.	Lowest and highest response time are the same, but intermediate rankings are different.
Accuracy per unit time	On both displays, BC has the highest APT, and SS chart has the lowest.	The lowest and highest APT are the same, but intermediate rankings are different.
Fixation rates	BC has the highest fixation rate on both displays.	BOR has the lowest fixation rate on VR display, while on 2D display AC has the lowest fixation rate.
3rd dimension	No similarities.	Movement of saccades along the z-axis is more than along the y-axis.
Number of unique sequences and revisit counts	BC has the lowest unique sequences and revisit counts on both displays.	BO has the highest unique sequences and revisit counts on VR display, while BOR has the highest unique sequences and revisit counts on 2D display.
Nominal variables	Colour variable has the highest accuracy and APT score on both displays.	Colour variable has the lowest fixation rate on VR display, while shape variable has the lowest fixation rate on 2D display.
Numerical variables	Size variable has the highest accuracy and APT score on both displays.	The opacity variable has the lowest fixation rate on VR display, while orientation has lowest fixation rate on 2D display.

Abbreviations: AC, area chart; BC, bar chart; BO, bar opacity chart; BOR, bar orientation chart; SO, shape opacity chart; SS, shape size chart; APT, accuracy (number of correct responses) per unit time.

dashboard starts with a bubble graph displaying fixation rate and duration while the participant was undertaking the task. Line graphs are used to investigate pupil diameter and saccadic intrusion, while multiple bar charts are used to analyze four EEG bands from different electrodes. We also used a pie chart that shows the average value of all electrodes. We can navigate between graphs by using the 'Switch Graph' button, and we can go back to the main menu by selecting the 'Main Menu' button. The 3D interface of the dashboard consists of multiple 3D bar charts to investigate EEG bands. We can also select and concentrate on any specific band by using the dropdown menu. We used 3D line graphs to analyze pupil diameter and saccadic intrusion, while a 3D radar chart was used to explore ocular parameters. The video of the flight simulation task is also displayed along

Figure 7.4 VR cognitive load dashboard.

with all graphs. The dashboard can be used beyond the flight simulator environment to visualize ocular data collected through any VR user study.

7.5 VISUALIZATION FOR USERS WITH VISUAL IMPAIRMENT

While we use data analysis and visualization in our everyday lives, from weather forecasting, judging product reviews during online shopping, checking financial trends for long-term planning and so on, but existing visualization techniques often turn inaccessible for low-vision and blind users. There are mainly three types of techniques for data visualization by blind users:

- The majority of systems use screen readers to articulate raw data or summary statistics
- Sonification of graphs with non-speech sound to indicate trends
- Raised paper or haptic interfaces

Providing interactivity along with screen reading is a challenging problem. Researchers at MIT developed systems for both navigation and authoring graphical content for a few types of graphs (Zong et al. 2022). One prototype, known as multiview, enabled individuals to use the up and down arrows to navigate between different levels of information (such as the chart title as the top level and the legend as the second level), and the right and left arrow keys to cycle through information on the same level (such as adjacent scatterplots). Another prototype, known as target, included the same arrow key navigation but also a drop-down menu of key chart locations so the user could quickly jump to an area of interest. The Umwelt (which means 'environment' in German) from the same research group (Zong et al. 2024) can enable blind and low-vision users to build customized, multimodal data representations without needing an initial visual chart. In such systems, researchers try to explore the spatial sense of blind users, and instead of developing a completely new system, they try to substitute visual sensory requirements of graphics through auditory or tactile means. Readers may check out the discussion in the context of spatial audio in Chapter 4.

During the COVID pandemic (2020–2022), we developed a new interactive visualization system to represent and compare the rate of spread of COVID across different countries over time (Biswas et al. 2020). We also developed an inclusive dashboard that was accessible to both users with visual and motor impairments. We integrated automatic speech recognition (ASR) and text-to-speech (TTS) features into our web application. We used the Web Speech API to implement these features. We included navigation features like scroll up and scroll down, along with the focus on selecting various countries through voice commands. Once we obtain access to a user's microphone, we classify the voice commands into three categories: a country name, a navigation command or an unrecognizable text. We have set the ASR recognition confidence threshold at 70%, and any predicted text with a lower confidence score would be rejected. On recognizing a country's name, we render the information on COVID-19 cases related to that country and use the TTS engine to read out the necessary information. This includes the total number of infections, the number of recovered cases and the number of deaths. This also includes information on new cases per day and new deaths per day across all three phases detected by our knee detection algorithm. On recognizing a navigation voice command, we simulate scroll events on our web application. On recognizing a command that is unrecognized, we use Levenshtein distance to suggest a country name based on the difference between our dictionary and the predicted text. The string in our dictionary with the least Levenshtein distance against the predicted text would be suggested to the user. A video demonstration can be found at https://youtu.be/lyvwK9enFtw. The system also had a large scrolling button and was tested with an eye gaze-controlled interface to make the system accessible to users with motor impairments.

7.6 QUICK FACTS

- **Visual variables:** Attributes of graphical elements that are used to encode and represent data in a visualization. These variables determine how data is visually communicated to the viewer and play a crucial role in making patterns, trends and relationships in the data understandable. There are eight primary visual variables: position, size, shape, brightness, colour, orientation, texture and motion
- **Graphical primitives:** Fundamental building blocks or basic elements used to construct visualizations. These are the simplest geometric shapes or forms that can be combined, manipulated or styled to create more complex visual representations of data. They are the atoms of a visualization, and all visualizations are composed of these basic components. Points, lines, circles and polygons are a few examples of graphical primitives
- **Data types:** These are the nature or characteristics of the data being visualized. Understanding data types is crucial because the type of data determines the appropriate visualization techniques and visual variables to use. Nominal, ordinal, quantitative and temporal data are examples of important data types
- **Gestalt principles:** A set of psychological principles that describe how humans naturally perceive and organize visual elements into meaningful patterns or groups. In the context of information visualization, Gestalt principles are essential for designing visualizations that are

intuitive, clear and easy to interpret. Proximity, similarity, continuity and closure are commonly used Gestalt principles
- **Interaction techniques**: Methods that allow users to dynamically explore, manipulate or filter data within a visualization. Interaction enhances the usability and depth of visualizations by enabling users to focus on specific data points, uncover hidden patterns or adjust the view to suit their needs. Zooming, panning, brushing and filtering are a few common interaction techniques

7.7 SUMMARY

Nowadays, a lot of decision-making is trusted to AI agents. However, there are many scenarios including medical, military, disaster response and product design where human sense of perception is either legally mandatory or trusted more than an AI agent. This chapter aims to give a brief introduction to the vast field of information visualization. Interacting and decision-making from large-scale data often lead to the development of novel interactive modalities or devices described in earlier chapters of this book. The chapter does not explicitly describe popular visualization tools like PowerBI or Tableau or various MATLAB packages; however, the user study described in the chapter can help to develop graphs or charts in two or three dimensions. Chapter 12 will revisit visualization in terms of existing standards in graph visualization.

Chapter 8

User study

8.1 INTRODUCTION

The previous chapters described how to develop technology. This chapter will explain how to evaluate technology. In previous chapters on AI (Chapter 3), computer vision (Chapter 4) and information retrieval (Chapter 5), we came across a few evaluation metrics like precision, recall and classification accuracy. However, when we evaluate human–machine interaction, the human element adds new challenges. Humans are neither fully deterministic nor random. We are (or at least hope to be) rational, meaning we try to do what we are supposed to do. However, we get 'used to' something or get 'bored' with something. Even after months of training, sometimes people can exceed expectations and similarly commit mistakes. In this chapter, we will learn to analyze interactions considering the variability in human activities. All of us are aware of collecting data and comparing means, but we will learn to compare means along with variance. The chapter is by no means a replacement for a standard statistics course, but we will mainly highlight how to design user studies, preprocess data, select appropriate statistical hypothesis tests and report results. The statistical tests are not specific to any statistical package, but once one prepares an appropriate data table, any software can be used to analyze results.

8.2 USABILITY EVALUATION

Checking conformance to well-established guidelines is the most direct approach to usability evaluation. Jakob Nielsen's ten usability guidelines are the most widely used heuristics. Similar guidelines, rather at an operational level, were published by the World Wide Web Consortium (W3C). There are plenty of automated tools available for checking conformance to WCAG (W3C Accessibility Guidelines), although such automated tools mostly check the syntax of webpages. Singh et al. (2024) compared multiple automated tools for checking conformance to WCAG. There are many features of UI/UX (user interface/user experience) design that require human

DOI: 10.1201/9781003646716-8

evaluators. In an ideal case, conformance to guidelines at a subjective level should be evaluated by multiple experts and consensus among experts' opinions should be reported (e.g., reporting Cronbach's alpha statistics). Besides heuristics, there are other methods of usability evaluation; one such method is cognitive walkthrough. In cognitive walkthrough, experts simulate users' interactions and walkthrough high-frequency to low-frequency tasks. The walkthrough can quickly identify errors/wrong assumptions in the structure/sequence of interfaces. There are other types of usability evaluation methods as follows:

- A pluralistic walkthrough uses group meetings where users, developers and usability experts step through a scenario, discussing each dialogue element
- A feature inspection lists sequences of features used to accomplish typical tasks, and checks for long sequences, cumbersome steps, steps that would not be natural for users to try and steps that require extensive knowledge/experience to assess a proposed feature set
- Consistency inspection designers, who represent multiple other projects, inspect an interface to see whether it does things in the same way as their own designs
- Standards inspection involves an expert on an interface standard inspecting the interface for compliance. Chapter 12 will discuss a few standards of interface design
- Formal inspection involves experts holding courtroom-style meetings with designers

Besides inspection or walkthroughs, usability is often evaluated through questionnaires or surveys. The IBM System Usability Scale, NASA Task Load Index (TLX) and Bedford Workload Scale are a few popular examples of standardized questionnaires used in usability evaluation with respect to users' acceptability of a system and self-reported cognitive load. While designing a bespoke questionnaire specific to a user study, negative questions or double questions should be avoided. Questions should also cover both positive and negative sides of a system. In Chapter 2, we also discussed the inclusive user model and Figure 2.6 shows an example of view of a person with visual and motor impairment. While simulation is not as popular as the previously described technique, a well-validated simulation can help designers visualize and measure the effect of impairment on design (Biswas 2010).

However, functionalities of a system need to be tested by end users even if the interface conforms to guidelines and is evaluated by experts. Self-reported assessments should also be complemented by objective data collected through controlled experiments. In the remaining sections of the chapter, we will discuss experiment design and data analysis.

8.3 EXPERIMENT DESIGN

In the present day, we often classify scientists as theoreticians and experimentalists, although applied research like our present topic of intelligent user interaction requires both theory and practice. However, the scenario was not the same always. Science is often credited to start with great philosophers who tried to answer and sometimes predict natural phenomena through logical reasoning. Prof Igor Novikov in his book on *The River of Time* noted:

> Galileo made a great many important discoveries in science that the reader undoubtedly knows about. However, the most important of these was his novel approach to natural sciences, his belief that to study nature one has first of all to set up carefully thought-out experiments. The world around us can only be understood by testing a hypothesis in experiments, by 'asking questions of Nature'. Here he parted ways sharply with Aristotle, who assumed that the world could be understood by purely logical reasoning. Galileo also believed that superficial observations not accompanied by thorough analysis of data can lead to wrong conclusions. Taken together, this was the beginning of the modern method of studying nature.

Following Galileo's footsteps, we had Sir Isaac Newton, Michael Faraday and a great many pioneers who taught designing experiments and validating theories.

In our present quest to evaluate intelligent user interface or interaction, we may broadly contextualize experiment design through the following bullet points, starting from generating a research or study question and analyzing results towards validation.

- **Generate research question**
  - Unstructured or semistructured interview
  - Observational methods
  - Coding and theorizing using qualitative data analysis techniques (discussed in the next section)
- **Postulate hypothesis**: A hypothesis is an if–then statement involving measurable terms. The word 'measurable' deserves special attention. For example, we cannot test whether an interface is good or bad, but we can test whether users can undertake a task quicker with less cognitive load than another alternative design
- **Define variables**: Variables are parameters that we can either control as part of experiment design or measure. The *independent variable* is the variable that is purposely changed. It is the manipulated variable as part of experiment design. Each independent variable can take one

of several possible values, known as the level of that variable. The *dependent variable* changes in response to the independent variable. It is the responding variable, like task completion time or number of errors. *Constant variables* are factors that are kept the same and not allowed to change. If the constant variables (like ambient light or noise, or posture of participants during a particular interaction) are changed and not accounted for, they can confound the results and conclusion from the experiment

- **Design experiment**
  - Sampling: Select a set of participants
  - Method: Design a study to collect data
  - Material: Get instruments
- **Measure**
  - Conduct experiment
  - Analyze results

It may be worth mentioning at this point that while we discuss experiment design, it often seems that our main intention is to prove the research hypothesis, but it is often more important to design an experiment for data collection rather than the immediate conclusion. The recent advancements in the field of computer vision and information retrieval are often attributed to the development of well-curated and annotated datasets that work as a benchmark for researchers. The first step of experiment design is deciding on a sampling criterion. For evaluating user interfaces, the sampling criteria can be decided based on intended end users like young users, elderly persons, users with a certain type of physical impairment, drivers and pilots. The conclusion from the study should be drawn upon this sampling criterion.

Experiment design can be classified based on different criteria, such as sampling strategy, duration of study and number of dependent variables. The experiment design involves identifying the independent variables, *levels of independent variables* and dependent variables. For example, independent variables can be the type of interface and font size. Each independent variable can have multiple levels, for example, interface type can be touchscreen, joystick and touchpad (three levels), while font type can have two levels (large font and small font). If we hypothesize about the trend of dependent variables (e.g., task completion time), the experiment design should be *longitudinal*, where we shall collect data for a prolonged duration of time and infer on the trend of data. Such designs are useful to analyze learning effects, or practice effects, or the effect of a particular environment (e.g., microgravity environment) on interaction. However, the majority of user studies on interface or interaction design are *cross-sectional*, meaning we collect data from different groups of users or on different design alternatives at a specific instance and use the results to compare across conditions.

Based on sampling criteria and research hypotheses, the experiment design can be matched pair, full factorial or fractional factorial. In *matched pair design*, one group of participants is matched with another group (male vs. female, expert vs. non-expert). In *full factorial design*, data is collected for all experimental conditions (all combinations of levels of independent variables), while in *fractional factorial design*, data is collected only for a subset of experimental conditions. When data is collected from the same participant in all experimental conditions, the experiment design is known as *repeated measures*. Repeated measures design requires fewer participants, but going through all conditions may be time-consuming for each participant, increasing the risk of fatigue or practice, known as *order effect*. Hence, the order of experimental conditions is randomized across participants. When we have access to a large number of participants, we may assign them to a subset of all experimental conditions, making sure we have an equal amount of data for each condition. The experiment designs will further be clarified in the next two chapters when we explain case studies.

The experiment design should also involve instruction to participants during data collection. For example, participants can be instructed to undertake a task as quickly as possible or as accurately as possible. The results will be different in each case. Similarly, participants can be instructed to fill out a post-test questionnaire based on their peak performance demand or average performance. The instructions to participants should be explicitly reported as part of the experimental design. Details on experimental design and statistical data processing can be found in Field (2009).

8.4 DATA ANALYSIS

Once the experimental setup is ready, we can proceed to data collection. Data are collected through logging data through a computing platform (e.g., computer, laptop, smartphone) or sensors (e.g., motion capture system, odometry for robotic agents). Next, we need to process the logged data and arrange it in the form of a table where columns designate different combinations of levels of independent variables, while the row represents participants or an individual task. It may be noted that most statistical tests will assume that each row is independent or not related to other rows.

Let us consider we are designing a user study to find the best combination of font size and spacing between icons on a computer screen (Figure 8.1). In such a scenario, the independent variables will be spacing and font size, and each of these variables may have the following levels:

- Spacing
    - Sparse
    - Medium
    - Dense

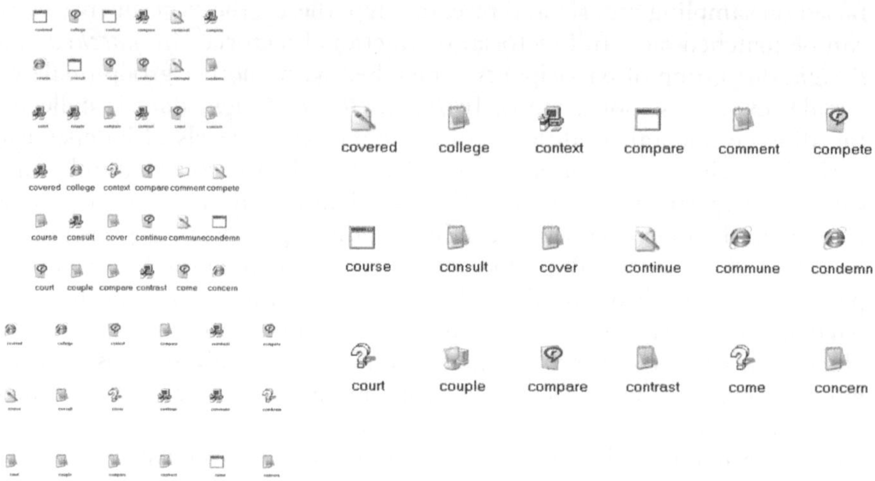

Figure 8.1 Icons with different spacing and font size of captions.

Participants	SparseSpacing SmallFont	SparseSpacing MediumFont	SparseSpacin gBigFont	MediumSpacing SmallFont	MediumSpacing MediumFont	MediumSpacing BigFont	DenseSpacing SmallFont	DenseSpacing MediumFont	DenseSpacing BigFont
P1									
P2									
P3									
P4									
P5									
.									
.									
.									

Figure 8.2 Sample data table for a study finding the best combination of icon spacing and font size.

- Font size
  - 10 pt
  - 14 pt
  - 20 pt

The data table should look like the table in Figure 8.2 where the cell values will contain values of dependent variables, that is, pointing and selection times, number of errors and NASA TLX values.

In another case, let us consider we are trying to find the best combination of line width and font size of text for axis labels and legends. In such a case, the independent variables will be line width and font size with the following levels (Figure 8.3):

- Line width: Two levels (thin and thick)
- Font size of legends and axes: Three levels (small, medium, large)

Once the data is properly organized, we need to screen the data before proceeding to statistical tests. Most statistical tests and also machine learning

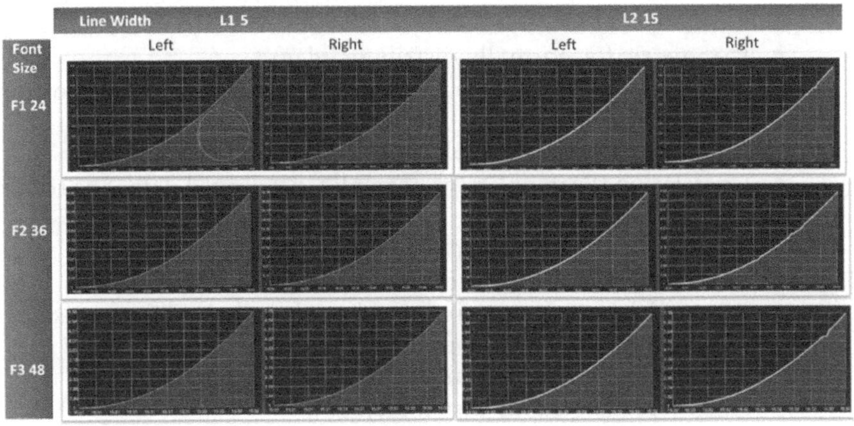

Figure 8.3 Graphs with different line widths and font sizes of axes labels.

models work based on assumptions about distribution of input data. The following bullet points highlight a few data screening techniques:

- **Outlier removal**: Outliers often change the distribution of data and pull the arithmetic mean towards outliers. Inner and outer fence can be used to detect outliers, and data points having values lower than the lower fence or higher than the upper fence can be removed from subsequent analysis.
  - Lower inner fence: Q1 – (1.5 × IQR)
  - Upper inner fence: Q3 + (1.5 × IQR)
  - Lower outer fence: Q1 – (3 × IQR)
  - Upper outer fence: Q3 + (3 × IQR)

Here IQR stands for interquartile range; it is calculated as the difference between the third and first quartile.

- **Unequal variance**: Many statistical tests like ANOVA (analysis of variance) assume that variances across different columns of data are nearly equal
- **Missing values**: If in a data table, any row of data has a missing column value, that row should be discarded
- **Data transformation**: We often run statistical tests on logarithmic values or ranks of original data to convert data distribution to Gaussian or normal distribution

Many statistical tests assume data is normally distributed and on a continuous scale. If data is not normally distributed or collected on a discrete scale (like in a Likert scale), we should use *non-parametric tests*. The following bullet points summarize different statistical tests:

- Data normally distributed – Kolmogorov–Smirnov (K-S) test, Shapiro–Wilk test
  - Parametric for normally distributed data
  - Non-parametric for data distribution that is not normal or on a continuous scale
- Relationship between two columns of data
  - Correlation (Pearson's *r* for normally distributed data; Spearman's ρ otherwise)
- Comparing means between two columns of data
  - A t-test for normally distributed data
  - Mann–Whitney U test (independent damples); Wilcoxon signed-rank test (dependent samples) otherwise
- More than two columns or more than one independent variable
  - ANOVA for normally distributed data
  - Kruskal–Wallis H test; Scheirer–Ray–Hare test (if more than two independent variables) otherwise
- Categorical data – Chi-square

Most statistical hypothesis tests work based on comparing means and variances across different columns of data. We use the t-test or ANOVA to calculate a ratio of variance between multiple groups (conditions) and within a group (condition), known as the *F-ratio*. As an experiment designer, we shall expect more variance between multiple groups (like interface A vs. interface B) rather than within a single group (among users of interface A). The value of this ratio and the number of groups or conditions and participants are used to estimate the probability of type I or false positive error known as *p-value*. The p-value indicates an estimate of the probability of not having a difference among groups. Figure 8.4 shows the typical output

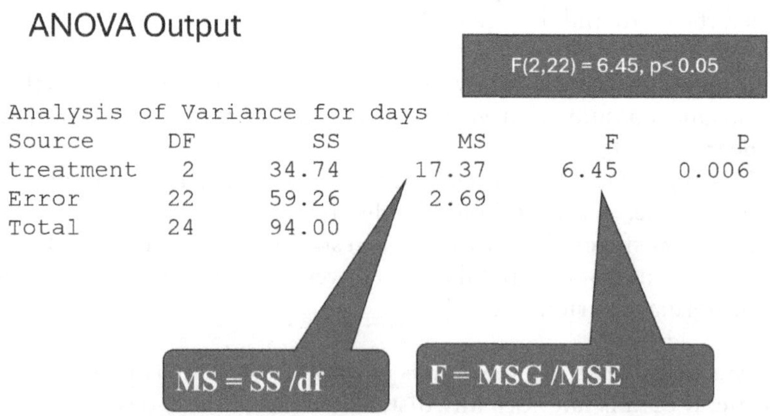

Figure 8.4 Example of output from an ANOVA.

from an ANOVA and corresponding reporting style. More examples of an ANOVA will be found in the next two chapters.

Finally, we shall briefly discuss qualitative data analysis techniques. As discussed in Section 8.2, qualitative data collection and analysis complement controlled experiments and quantitative results. We may collect data through interviews or observational methods or other means before, during or after controlled experiments for generating research hypotheses, capturing users' subjective feelings and studying user acceptance of the final product of design, respectively. Qualitative data analysis techniques can use similar techniques of text preprocessing and keyword analysis as discussed in the chapter on information retrieval. However, the process is not completely automated; rather, an expert goes through transcripts and manually annotates the transcript with codes. Later, the codes are further analyzed and combined into interrelated themes. There are standard procedures like the grounded theory-based approach for this thematic analysis procedure; however, the process often involves manual interpretation of raw text, codes and themes. We shall present results on a thematic analysis of user acceptance of an interactive head-up display of a car in Chapter 10.

8.5 QUICK FACTS

- **t-test**: The term *t-statistics* is abbreviated from 'hypothesis test statistic'. Student's t-test is a statistical test used to test whether the difference between the responses of two groups is statistically significant
- **Analysis of variance (ANOVA)**: ANOVA is a family of statistical methods used to compare the means of two or more groups by analyzing variance. Specifically, ANOVA compares the amount of variation between the group means to the amount of variation within each group. If the between-group variation is substantially larger than the within-group variation, it suggests that the group means are likely different
- **Statistical significance**: In statistical hypothesis testing, statistical significance measures the probability of the null hypothesis being true compared to the acceptable level of uncertainty regarding the true answer. A study's defined significance level, denoted by α, is the probability of the study rejecting the null hypothesis, given that the null hypothesis is true
- **Type I and type II errors**: In statistical hypothesis testing, a type I error, or a false positive, is the erroneous rejection of a true null hypothesis. A type II error, or a false negative, is the erroneous failure to bring about appropriate rejection of a false null hypothesis
- **p-value**: It is an estimate of the probability of a type I error

8.6 SUMMARY

In summary, this chapter aims to explain designing user studies and analyzing results. We briefly discussed different types of experimental designs and selecting appropriate statistical tests. It may be noted that although we spent more time on explaining statistical tests, more emphasis should be given to experimental design and data collection. A well-collected data source can be utilized by researchers over years, while an interesting result from an improperly collected data sample has no use. In the upcoming two chapters on human–robot interaction and case studies, we will explain user study design for different applications.

Chapter 9

Human–robot interaction

9.1 INTRODUCTION

The word *robot* was first used to describe artificial automata in a Czech play in 1920 and literally meant 'forced labour'. From the inception of the term and recent progress in the field of AI, society has often panicked society about the effect of robots in terms of job loss, machine takeover and a loss for humanity. However, robots have kept on making inroads in multiple segments of society, from industrial automation to precision surgery. This chapter will discuss human–robot interaction (HRI) for three different applications: collaborative robots (cobots) for industrial applications, social robots and assistive robots. We will briefly discuss the main goals of each application, followed by a case study on developing safe human–robot interaction for persons with severe speech and motor impairments. A second case study will briefly describe the building blocks of a LLM-controlled semi-autonomous rover, where the destination will be given by a human through natural language or even an image, and the robot autonomously navigates to the desired destination, avoiding obstacles. The chapter will emphasize applications of intelligent user interfaces for semi-autonomous control of robotic agents.

9.2 COLLABORATIVE ROBOT (COBOT)

Cobots were invented in 1996 by J. Edward Colgate and Michael Peshkin, professors at Northwestern University. Their US patent titled 'Cobots' describes 'an apparatus and method for direct physical interaction between a person and a general purpose manipulator controlled by a computer' (ISO/TS 15066:2016). The invention resulted from a 1994 General Motors initiative led by Prasad Akella of the GM Robotics Center and a 1995 General Motors Foundation research grant intended to find a way to make robots or robot-like equipment safe enough to team with people. A human–robot collaborative situation can be classified into:

DOI: 10.1201/9781003646716-9

- **Coexistence**: Humans and robots work alongside each other without a fence, but with no shared workspace
- **Sequential collaboration**: Humans and robots are active in a shared workspace, but their motions are sequential; they do not work on a part at the same time
- **Cooperation**: Robots and humans work on the same part at the same time, with both in motion
- **Responsive collaboration**: The robot responds in real time to the movement of the human worker

The ISO 10218 standard (2016) precisely describes a formula to keep a minimum protective separation distance between the cobot and the operator's body parts, considering the stopping distance of the robotic system, and velocities of the robot and operator, including uncertainties in the positions of both the operator and robotic parts. A few applications for cobots in industry are as follows:

- Pick and place
  - Manual pick-and-place tasks are often the most repetitive and mundane tasks in a production environment, with the dull nature of the task often leading to mistakes or decreased efficiency if workers are not frequently rotated
  - These are generally some of the least enjoyable tasks that can also entail a risk of repetitive strain or other injury if the object is of substantial weight
- Machine tending
  - The cost of a cobot is usually significantly lower than the machine it is tending. By allowing the machine to run 24 hours of back-to-back cycles, this will likely pay for the cobot tending it in a very short period
- Packaging and palletizing
  - The benefits of automating a packaging task are similar to generic pick and place – increased productivity over multiple shifts
- Process tasks (gluing, dispensing or welding)
  - In a process task, whether it's gluing, dispensing or welding, the key details are the same: the robot moves a tool through a fixed path while the tool interacts with the workpiece
  - In each of these process tasks, it takes a significant amount of time to train a new employee to be able to control the numerous variables required to attain an excellent quality finish
  - If this control can instead be copied directly from one robot to another, it becomes a considerably more straightforward process
- Finishing tasks (polishing, grinding or deburring)

    - As with a process task, the possibility for sharp productivity and quality improvements is immediately clear when automating a finishing task with a cobot
- Quality inspection
    - A machine vision system will usually have a faster, more consistent output than a person inspecting a product, leading to improved quality and productivity

9.3 ASSISTIVE AND SOCIAL ROBOTS

From industrial robots, we will next discuss the use of human–robot interaction in household applications. According to a World Health Organization (WHO) report on disability, 15% of the world's population lives with some form of disability. A major portion of this population suffers from severe disabilities, rendering them unable to talk, walk, eat and interact naturally and independently with their environments. Prosthetics and powered wheelchairs are common assistive technologies, which are not far from robotic manipulators or moving bots. Advanced sensors and intelligent algorithms enable robotic arms to be used for independently feeding persons with disabilities by carefully monitoring humans, identifying food items on the table, planning their trajectory and controlling the orientation of the end effector. Robot exoskeletons help persons with limited mobility to walk and move around by inducing calculated electrical impulses in the muscles, amplifying the forces generated and maintaining the upright posture for walking momentum. Robots equipped with cameras and ultrasonic sensors help persons with vision impairments to safely navigate through everyday environments. Repositioning and transferring robots help caregivers in moving, shifting and lifting elderly persons from wheelchairs and beds and changing positions.

Personal and social robots help by providing companionship, reminding medication schedules, telling jokes, responding with facial expressions and soothing sounds when petting or touching them. Social robots are different from assistive robots in terms of providing social support by following social behaviours and rules attached to their role, mainly to elderly persons or persons with cognitive impairment. Unlike assistive robots, social robots take the shape of a living creature, like a humanoid or a popular pet, to enhance social bonding with users. Human–robot interaction with social robots is often based on affect or emotional state recognition of users or helping users with cognitive impairment, such as autism, to improve skills to interpret emotional states.

9.4 CASE STUDY ON XR-BASED HRI

All of us get fortunate in life in seeing children growing up. As a child starts to crawl, speak and walk, it brings joy to the parents and relatives.

However, children with a different range of abilities often follow a different development path than their able-bodied counterparts. People with severe speech and motor impairment (SSMI) often find it difficult to manipulate physical objects due to spasticity and have familiarity with eye-pointing-based communication. Such persons cannot control their muscles voluntarily. This lack of control may arise from an accident, disease or genetic disorder, limiting the person's mobility to a wheelchair with constant care and assistance needed. In such a situation, we propose a safe human–robot interaction system involving a tabletop, low-speed, collaborative and multimodal robotic arm. The arm can be controlled with the user's eye gaze for picking and placing, drawing, painting, playing with toys and so on. All activities happen in proximity to the personal space of a human. Hence, the safety of the user and the caregiver, in and around the robotic arm, is given top priority in HRI system design. The robotic arm needs to avoid any kind of collision or impact with the hand.

We developed an eye-gaze controlled human–robot interface for users with SSMI to help in their rehabilitation process by undertaking different object manipulation tasks like painting and fabric printing. In particular, the fabric printing task involves cooperation between an able-bodied person and a person with SSMI. The person with SSMI picks up a string about 6 cm in length and drops it on a powdered dye with his hand. He cannot pick it up himself from the dye. The able-bodied person picks up the coloured string and rubs it on a piece of paper. The whole process is repeated a few times with different coloured powder dye. It may be noted that the person with SSMI only chooses the colour but cannot put the colour in his or her chosen position on the piece of paper. We divided the fabric printing task into two activities: a pick-and-drop task and then a stamping task. Subsequent simulation and user studies investigated this pick-and-drop task; the whole activity including the stamping task can be viewed at https://youtu.be/PIv8v46Gtis?t=54. The tasks are chosen as representative of primitive pointing and selection made using eye-gaze control through an augmented video see-through interface by users with severe speech and motor impairment to operate a tabletop robotic arm. A gaze-controlled 5 degrees of freedom robotic system able to manipulate a payload of 330 grams was developed to enhance the recreating activities of the user with SSMI.

The gaze-controlled user interface (UI) includes:

- **Video see-through (VST) user interface**: A VST interface can be deployed without a head-mounted rendering system and can be integrated with an eye-gaze tracker and robotic manipulator. The proposed user interface is shown in Figure 9.1d. A user interface was rendered on the augmented reality (AR) VST display enabling users to operate the display by aiming their eye gaze on screen elements,

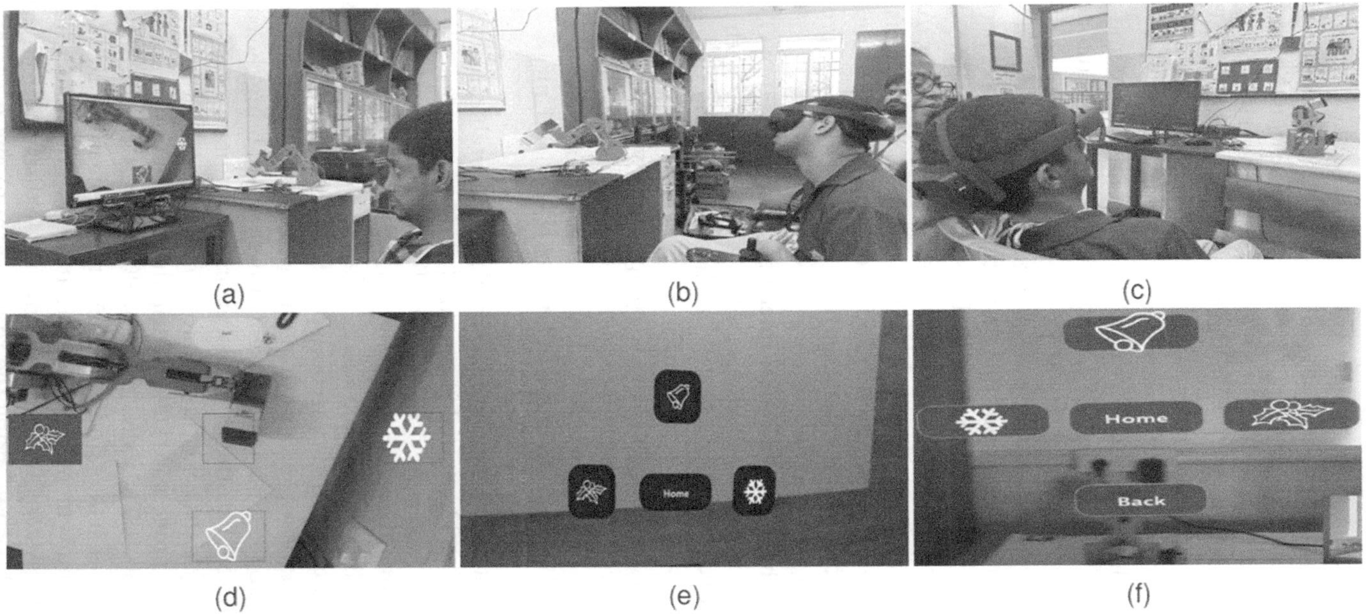

Figure 9.1 (a) SSMI user using Tobii eye tracking to operate the stamp printing robot. (b) SSMI user utilizing AjnaLens headset (VPT) to operate the stamp printing robot. (c) SSMI user using HoloLens 2 headset (OST) to operate the stamp printing robot. (d) AR video see-through user interface. (e) Mixed reality user interface integration with AjnaLens. (f) Mixed reality user interface in HoloLens 2.

as illustrated in Figure 9.1a. All virtual screen elements adapted their contrast based on ambient light conditions, and the display matches the dimensions of the physical world through an offset correction. As the user selects a screen element, a command is sent to the robotic arm to undertake an action fulfilling the users' intention
- **Video pass-through (VPT) user interface**: A mixed reality VPT interface was developed for the AjnaLens Pro HMD, a device with a 4560 × 2280 pixel resolution, 90 Hz refresh rate and a field of view between 95 and 105 degrees. This device provided an immersive and realistic experience. The proposed VPT interface was designed and deployed, ensuring smooth gaze interaction as shown in Figure 9.1b. A key feature of the AjnaLens Pro was its lightweight design (400 grams), which made it more comfortable than other HMDs. Users performing the stamping task through the AjnaLens are shown in Figure 9.1. The deployment process involved integrating the UI with the AjnaLens Pro's software development kit (SDK). Once the scene was set up, the project settings for AjnaLens Pro compatibility were configured. Testing was conducted to ensure that the user interface provided a seamless and immersive experience
- **Optical see-through (OST) user interface**: A mixed reality OST interface for the Microsoft HoloLens 2 HMD was designed with a Mixed Reality Toolkit (MRTK). The device has a resolution of 1440×936 per eye and a 60 Hz refresh rate. The user interface was designed as shown in Figure 9.1f. MRTK components were fully integrated into the project. The scene was set up with essential MRTK features including spatial awareness and gaze input. The project was built through the Universal Windows Platform (UWP) in Unity

The robotic manipulator is supposed to be placed on a table, has a working envelope of 60 cm × 32 cm and is operated through a video see-through interface described in the previous section. The user will be at least 65 cm away from the manipulator; however, they may put their forearm and palm on the table, and the robot should not be allowed to come near these body parts. We have integrated the MediaPipe Hands system to create a safe and collaborative tabletop robotic arm operation. Once the hand region is detected, we developed a navigation algorithm that will bring the end effector of the robotic arm to the target location by maintaining a minimum distance from body parts. We propose to model the navigation problem as a Markov decision process (MDP) and use the value iteration algorithm to find the optimal policy. The value iteration algorithm is modified to remember past positions of hands and can adapt the navigation path based on uncertainty in the movement of the robotic arm. A set of simulation studies confirmed that the system always maintains a safe distance from the hand, and users with SSMI could bring the robotic arm to any designated point within its working envelope in less than 3 minutes. A user study involving 13 participants and screen-based VST showed that all could complete the trial; the maximum duration for the first attempt was 3.4 minutes, while for the second attempt, it was 2.6 minutes. The average separation distance was 25 cm between the user's palm and robot's end effector, and it never fell below 13 cm. We undertook a between-subject ANOVA (Trial number × Participant type) among the task completion times. We found:

- Significant main effect of trial $F(1, 11) = 8.648$, $p < 0.05$, $\eta^2 = 0.44$
- Significant main effect of participant type $F(1, 11) = 106.16$, $p < 0.05$, $\eta^2 = 0.906$

We did not find any significant interaction effect between trial number and participant type. Figure 9.2 shows the task completion times for two different types of participants and trial number; it may be noted that for both groups of users, their task completion time significantly decreases at $p < 0.05$ (Himanshu 2025) in paired t-tests.

A second study was conducted to evaluate three interfaces: VST, VPT and OST, each employing different eye tracking. The average stamping time and the number of stamps across different interfaces are shown in Figure 9.3. The average stamping time for VST was the lowest at 28.45 seconds, followed by OST and VPT at 37.97 seconds and 40.31 seconds, respectively. The average number of stamps for VST, VPT and OST were 7.36, 6.64 and 6.27, respectively. The system demonstrated an average stamp count of at least 6.27. The system shows users with SSMI were already familiar with the VST system from the pilot study, resulting in faster stamping times and the highest number of stamps with VST compared to the other interfaces. The average number of stamps using VPT and OST was similar. The shorter stamping time for OST indicates that users with SSMI completed the task more quickly with the HoloLens than with the AjnaLens.

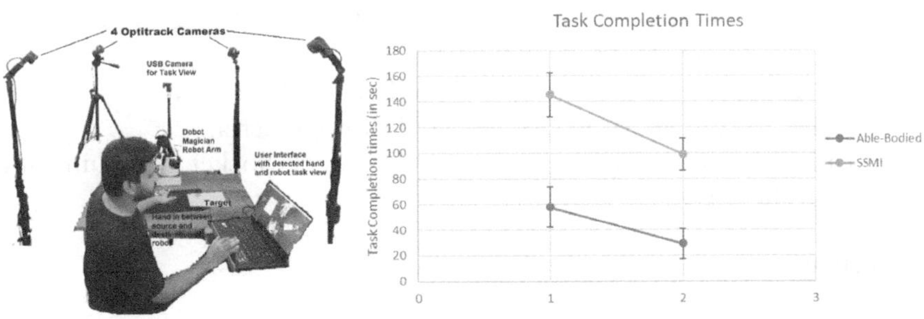

Figure 9.2 Experimental setup and task completion times during user study. (Left) The setup with the Optitrack Motion Capture System. (Right) Improvement in human–robot interaction of both groups of users.

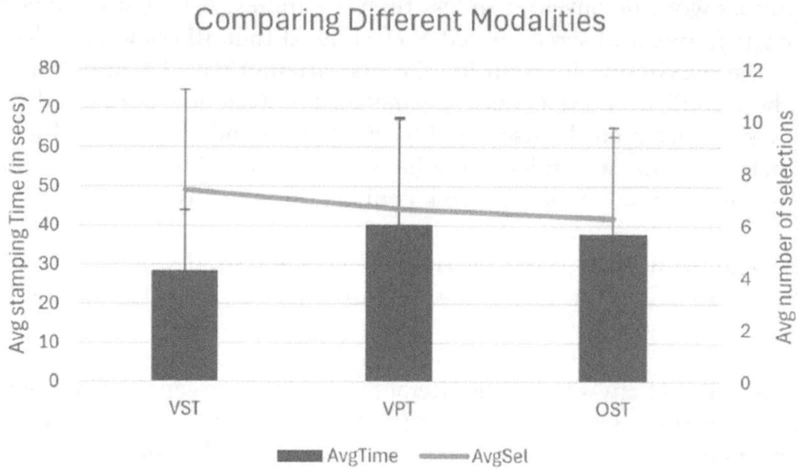

Figure 9.3 Comparison of stamping time and number of stamps for different interaction interfaces. The primary y-axis shows average stamping time, while the secondary y-axis shows average stamp count, and the x-axis shows different interfaces.

Details of these studies can be found in Sharma et al. (2022) and Himanshu et al. (2025).

Next, we will present a quick guide outlining the process of creating an interactive system in Unity where a user can control a robot through eye gaze using an extended reality (XR) headset with eye-tracking capabilities.

9.4.1 Creating an interactive system in Unity

Requirements

- Unity (2021 LTS or later recommended)
- XR plugin (OpenXR or Mixed Reality Toolkit [MRTK] for HoloLens)
- Eye-tracking SDK (Tobii XR SDK)
- XR headset with eye tracking (HoloLens 2, AjnaLens)
- Robot (Arduino-based or connected through Serial)
- Unity communication plugin (for Serial or Socket communication with the robot)

Objective

- Use eye gaze to select virtual targets in the scene
- Trigger robot commands based on gaze-based selection
- Provide visual feedback (e.g., gaze cursor or highlights) in XR

Step 1: Unity project setup

- Create a new 3D Unity Project
- Install XR Plug-in Management
  1. Navigate to Edit → Project Settings → XR Plug-in Management
  2. Add OpenXR for HoloLens 2
- HoloLens 2
  1. Import and configure Mixed Reality Toolkit (MRTK) from GitHub or Unity Package Manager
  2. Use the MRTK Configuration Wizard to set up the input system and camera profile

Step 2: UI development in Unity

See Figure 9.4.
Add a Canvas to the scene:

- Set Render Mode to Screen Space – Overlay or World Space (for XR integration)

Add UI elements:

- Buttons for commands like Start Robot, Reset, Move to Position 1, etc
- Text (or TMP Text) for instructions and labels

Add interactivity:

- Attach C# scripts to handle UI events (OnClick(), OnValueChanged()

Step 3: Eye tracking configuration (for MRTK + HoloLens 2)

- Import Microsoft.MixedReality.Toolkit.Input module
- In the Input System Profile, enable Eye Tracking
- Add the Eye Gaze Provider to the XR camera object (Figure 9.5)

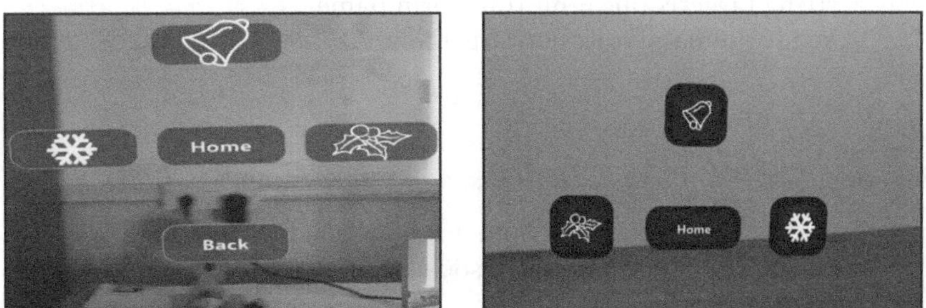

Figure 9.4 User interface for task selection.

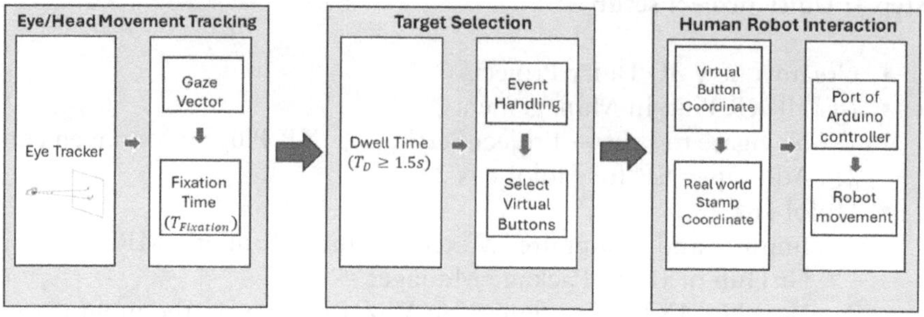

Figure 9.5 Target selection through eye gaze. A set of steps used to implement a gaze-controlled human–robot interaction system.

Step 4: Detect gaze targets

Initialization:

- Define gazeRay from the XR headset or eye direction
- Set gazeLayer to include interactive objects only
- Set maxRayDistance to limit how far gaze raycasts (Figure 9.6)

Logic:

```
//Step 1: Get the direction in which the user is currently looking
Vector3 eyeDirection = EyeTrackingProvider.GazeDirection;
// Step 2: Perform a raycast to detect if the gaze hits any target on the
specified layer
RaycastHit hit;
if (Physics.Raycast(gazeOrigin, eyeDirection, out hit, maxRayDistance,
gazeLayer)) {
// Step 3: Provide visual feedback, such as highlighting the object
   ShowGazeFeedback(hit.transform); // e.g., highlight
   if (UserTriggersSelection()) {
      string targetName = hit.transform.name;
      SendCommandToRobot(targetName);
   }
}
```

Function: *SendCommandToRobot(string target Name)*

- Translate the object name into a robot-friendly command.
- Send the command via serial, socket.

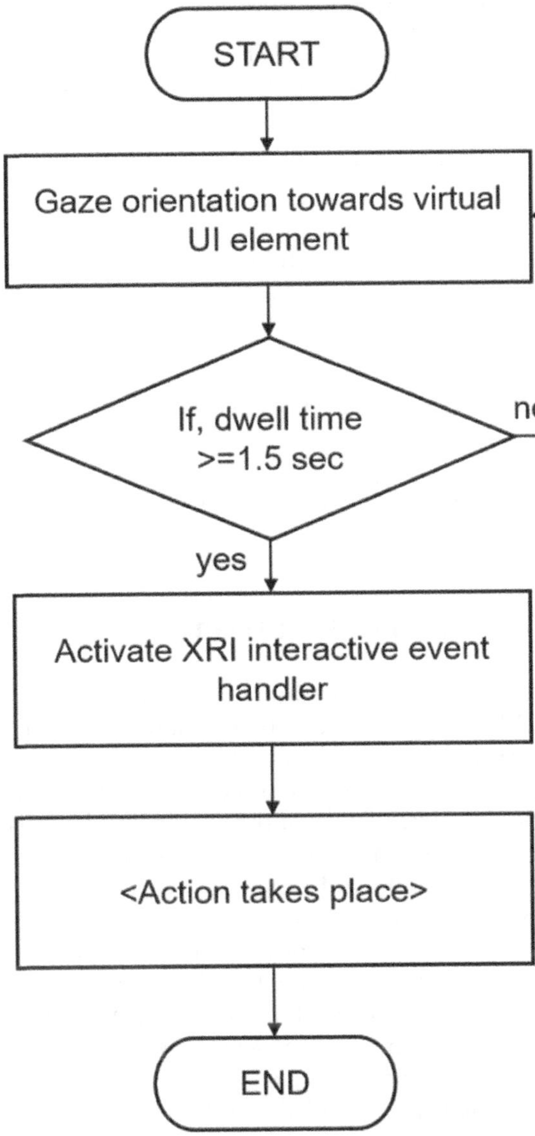

Figure 9.6 The process of target selection through eye gaze.

Step 5: Robot command communication

- Use Serial Communication, Socket
- Bind UI events to send commands to the robot based on user input
- Serial (USB → Arduino)

```
System.IO.Ports;
port = new SerialPort("COMx", 9600);
void Start () => port.Open();
void SendToRobot(string command) =>
port.WriteLine(command);
```

Step 6: Build and deploy

- Build the Unity app for your XR device
- Test for:
  - Gaze accuracy and responsiveness
  - Successful command transmission to the robot
- Ensure robot reacts properly based on user interaction

9.5 CASE STUDY ON LLM-BASED HRI

This section presents another case study to operate a rover through large language model (LLM). LLMs offer a powerful tool for understanding natural language instructions for multiple domains. However, for the robot control domain, often the richness of a natural language command is lost during tokenization to a specific set of robot commands. This project interprets natural language commands for cyber-physical systems based on their availability, relative positions and range of abilities. We present a multimodal human–robot interface designed to enable users to interact naturally with a robot using text or images. The system described next assumes three locations named Base Station, Station 1 and Station 2, and autonomously navigates to one of the stations based on user input. We can easily add more stations without changing the internal workings of the system. A user can describe a task (e.g., 'go to your charging station' or submit a picture of tools or speak an instruction), and the system intelligently interprets the input and selects the most relevant station (e.g., Station 1, Station 2 or Base Station). This decision is then sent to the robot system via socket communication. The system is composed of a web app frontend, a Python-based backend using Ollama models, and a real-time socket communication layer. A video demonstration can be found at https://youtu.be/AIELvvTQTnU.

1. User input (text, image or audio)

Text input: The user can enter commands or task descriptions in a text box (e.g., 'Send the wrench to Station 1'). This input is directly serialized and forwarded to the backend.

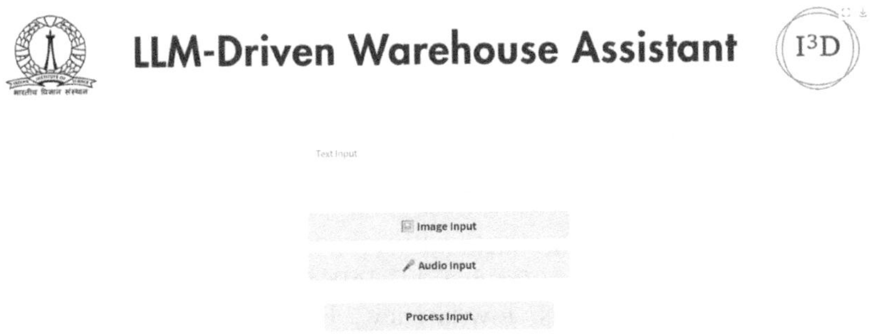

Figure 9.7 Web app frontend for LLM integration.

Image input: The user can upload or select an image (e.g., tools or components). Images are displayed in the interface and then converted to base64 strings before transmission.

Audio input: Users may also upload or record audio clips describing a task. Upon upload, the Whisper ASR model (base.en) is used to transcribe the audio to text. Whisper is run locally in the frontend to preserve responsiveness and reduce latency (Figure 9.7).

All three types of inputs are unified into a consistent JSON payload structure and transmitted to the backend via a TCP socket.

2. Sending input to the backend

Inputs are serialized as newline-delimited JSON strings, allowing efficient and stateless transmission over a persistent TCP socket connection.

Example payloads:

- Text message:

```
{
 "type": "text",
 "data": "Go to your charging station"
}
```

- Image input:

```
{
 "type": "image",
 "data": "<base64-encoded image>"
}
```

- Transcribed audio:

```
{
 "type": "text",
 "data": "Do all the hazardous materials have warning
signs on them?"
}
```

All payloads are sent to the backend server listening on localhost:12346 via a standard Python socket (socket.AF_INET, socket.SOCK_STREAM). This architecture supports low-latency, bidirectional communication between the frontend and the decision-making backend (Figure 9.8).

3. Backend processing

The backend is a Python-based server running on a local GPU-enabled machine. It leverages models served via Ollama to process both text and image inputs and determine relevant robotic actions. The processing workflow varies based on input type.

- Text input
  - Text inputs are processed using the **LLaMA 3.1 (8B)** model via the Ollama API
  - Prompts are dynamically constructed to perform reasoning and select the most relevant location for a tool
  - The model is constrained to choose one option only from:
    - Station 1 (for tools like hammers, screwdrivers, etc.)
    - Station 2 (for hazardous or dangerous items)
    - Base Station (for base/home/charging-related inputs)

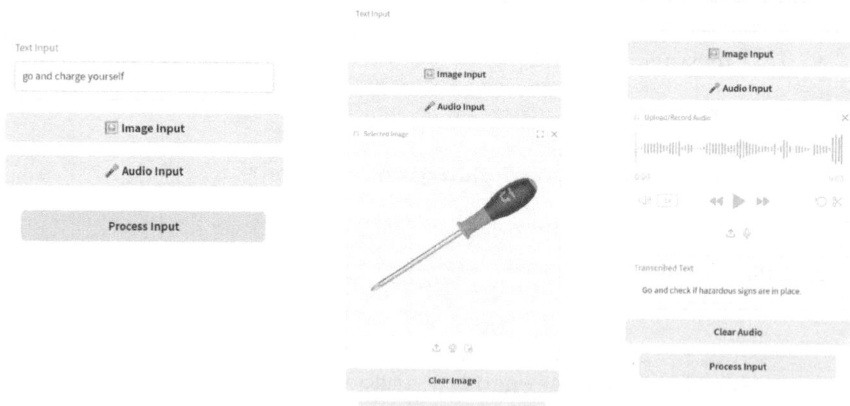

Figure 9.8 Text, image and audio inputs that can be mapped to physical location for a rover.

```
def interpret_task(self, user_sentence):
       try:
            response = ollama.chat(
                model="llama3.1:8b-instruct-fp16",
                messages=[
                    {
                        "role": "user",
                        "content": f"""
You are an assistant trained to identify the most relevant
location based on the input sentence. Choose only one of the
following options:
Station 1: for tools (e.g., hammer, screwdriver).
Station 2: for hazardous materials or anything involving
danger or caution.
Base Station: for home position, base station, or charging
station references.
Respond with exactly one of the options above, or 'None' if
no location applies.
Sentence: "{user_sentence}"
"""
                    },
                ]
            )
            response_content = response.get("message", {}).
get("content", "").strip()
```

- Image input
    - Image inputs are processed using the llama3.2-vision model via Ollama
    - Images are temporarily saved and sent to the model with a concise prompt asking for object names only, stripped of any adjectives or additional context
    - Detected objects are then passed through the same LLaMA text reasoning model to determine the appropriate location
    - This enables multimodal understanding (vision + reasoning) in a two-step pipeline

```
def object_detection_ollama(image):
    try:
        temp_image_path = image_path # image as numpy array
        pil_image = Image.fromarray(cv2.cvtColor(image,
cv2.COLOR_BGR2RGB))
        pil_image.save(temp_image_path)

        concise_prompt = "List only the names of the
objects present in this image, separated by commas. Do not
```

```
include any additional information, descriptions,
adjectives, or possessive terms."
        # Send the image to Ollama's vision model with the
concise prompt
        response = ollama.chat(
            model='llama3.2-vision',
            messages=[{
                'role': 'user',
                'content': concise_prompt,
                'images': [temp_image_path]
            }]
        )
        # Clean up the temporary file
        os.remove(temp_image_path)

        # Process the response
        if 'message' in response and 'content' in
response['message']:
            detected_objects =
response['message']['content'].strip().rstrip('.!?')
            return detected_objects
        else:
            return "Ollama did not return a valid
response."

    except Exception as e:
        return f"Error during object detection with
Ollama: {e}"
```

Audio processing

- Audio is handled via Whisper, which transcribes the waveform into a textual prompt
- The resulting transcript is passed into the same text-processing pipeline described earlier

Backend decision flow

1. Receive payload (text/image)
2. Identify type (text, image, etc.)
   Call appropriate model:
   - ollama.run("llama3.2-vision", prompt, image) for vision
   - ollama.run("llama3.1:8b", prompt) for pure text
3. Parse output, generate structured response

4. Sending response to the frontend

Once processing is complete:

- The backend generates a structured JSON response
- This response is sent back via the same socket connection to the frontend

Backend response:


```
{
  "type": "text",
  "data": "Relevant Station: Base Station"
}
```


- The system is optimized for single-turn interactions

5. Displaying backend response in the frontend

The frontend maintains a lightweight listener that polls the backend every 2 seconds for updates using the same socket connection.

- Upon receiving a backend message, the frontend displays it using a toast notification (transient popup)
- This ensures the user is kept informed without cluttering the interface (Figure 9.9)

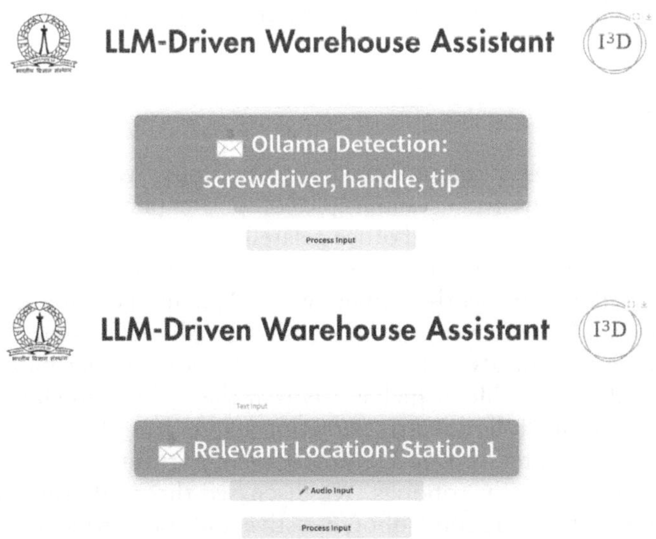

Figure 9.9 Messages from the backend of LLM.

9.6 QUICK FACTS

- **Industrial robot/robot**: It is an automatically controlled, reprogrammable multipurpose manipulator(s), programmable in three or more axes, which can be either fixed in place or fixed to a mobile platform for use in automation applications in an industrial environment
- The industrial robot includes:
  - The manipulator(s), including robot actuators controlled by the robot control
  - The robot control
  - The means to teach or program the robot, including any communications interface (hardware and software)
- **Cobot**: An apparatus and method for direct physical interaction between a person and a general-purpose manipulator controlled by a computer. The apparatus, known as a collaborative robot or 'cobot', may take a number of configurations common to conventional robots
- **Separation distance**: It is the shortest permissible distance between any moving hazardous part of the robot application and any operator
- **SLAM**: Simultaneous localization and mapping (SLAM) is the computational problem of constructing or updating a map of an unknown environment while simultaneously keeping track of an agent's location within it
- **Social robot**: It is an autonomous robot that interacts and communicates with humans or other autonomous physical agents by following social behaviours and rules attached to its role. Like other robots, a social robot is physically embodied (avatars or onscreen synthetic social characters are not embodied and thus distinct)

9.7 SUMMARY

From the mid-1980s era of personal computers, now we are in a time where each individual owns and uses multiple computing platforms in the form of handheld, wearable, home and office-related devices. The next big technical trend may be where a single operator is operating multiple cyber-physical systems just like we presently manage WhatsApp, mail client and Netflix all together on multiple devices, often simultaneously. Just as we do not read user manuals or tutorials on how to use social media or microblogging sites, robots can also be usable in such a way that they will be intelligent enough to understand and follow instructions in a safe and efficient manner. In summary, this chapter aims to highlight one important application of intelligent user interfaces for robotics. We discussed three different approaches of human–robot interaction (cobot, assistive and social robots) followed by two case studies. The case studies throw pointers to upcoming use cases on naturalistic human–robot interaction for societal causes.

Chapter 10

Case studies on cockpit design

10.1 INTRODUCTION

Previous chapters described many theoretical aspects of designing intelligent user interaction with a few examples of mixed reality, information retrieval and visualization systems. Chapter 9 discussed applications of human–robot interaction. This chapter will describe three case studies on cockpit design, where design decisions often play a role between life and death. There are numerous examples where a seemingly minor error by an operator led to a fatal disaster. Similarly, there are scopes for integrating new technology that can bring strategic advantage to a product, company or even a nation.

We will discuss three case studies on incorporating new technology or evaluating existing human–machine interfaces (HMIs) in a cockpit. We will start with developing an interactive gaze-controlled head-up display (HUD) for cars with an in-built distraction detection feature. Next, we will move to the aviation domain and describe a case study on developing a gaze-controlled display in different configurations. Our last case study compares existing HMIs for manned spacecraft cockpits. For each case study, we briefly describe contemporary research challenges, proposed solutions and results from user studies involving end users. Together, the three case studies will summarize the contents of our previous chapters on alternate modalities, human factors, AI/ML and usability evaluation. Readers will get an ensemble view of the design, development and evaluation of intelligent user interfaces.

10.2 INTERACTIVE HUD FOR CARS

Modern infotainment systems in automobiles facilitate driving at the cost of secondary tasks in addition to the primary task of driving. These secondary tasks have a considerable chance of distracting a driver from their primary driving task, thereby reducing safety or increasing cognitive workload. This section presents an intelligent interactive HUD on the windscreen of the

DOI: 10.1201/9781003646716-10

driver that does not require them to take their eyes off the road while undertaking secondary tasks such as playing music, operating vent controls or watching the navigation map. The interactive HUD (IHUD) allows interaction in the form of pointing and selecting just like traditional graphical user interfaces, while tracking operators' eye gaze or finger movements. Additionally, the system can also estimate drivers' cognitive load and distraction level. User studies show the system improves driving performance in terms of mean deviation from lane in an ISO 26022 lane-changing task compared to a touchscreen system, and participants can undertake ISO 9241 pointing tasks in less than 2 seconds on average inside a car.

The proposed system consists of hardware and software parts described next, and Figure 10.1 is a picture of the setup inside a car. The system can be deployed in two ways:

1. Using reflection from a screen with high brightness on a semitransparent sheet placed on the windscreen (Figure 10.1a)
2. Overlaying controls on the video feed of a tablet computer (Figure 10.1b)

We have developed an artificial neural network-based eye-gaze calibration program to move a cursor on the reflected display in the windscreen. We also developed a finger movement tracking program using a LeapMotion controller. The gaze-controlled interface is explained in Figure 10.2. We conducted five user studies as follows:

1. Study involving a simulator comparing the interactive HUD with and without hotspots and touchscreen system for an existing JLR dashboard
2. Study inside a running vehicle with the eye-gaze tracking system set up for passengers

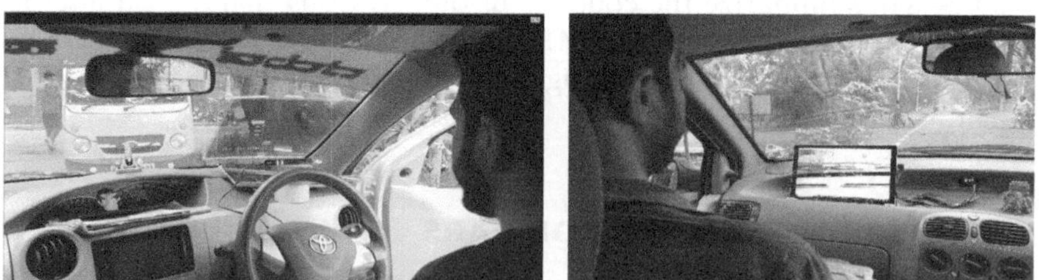

a. With reflected screen (Optical See Through Display) b. Using tablet computer (Video See Through Display)

Figure 10.1 Novel interactive head-up display system for cars. (a) A configuration where a tablet display is reflected on the windscreen. (b) We configured it as a video see-through display where interactive elements were rendered on top of the live video feed from the back camera.

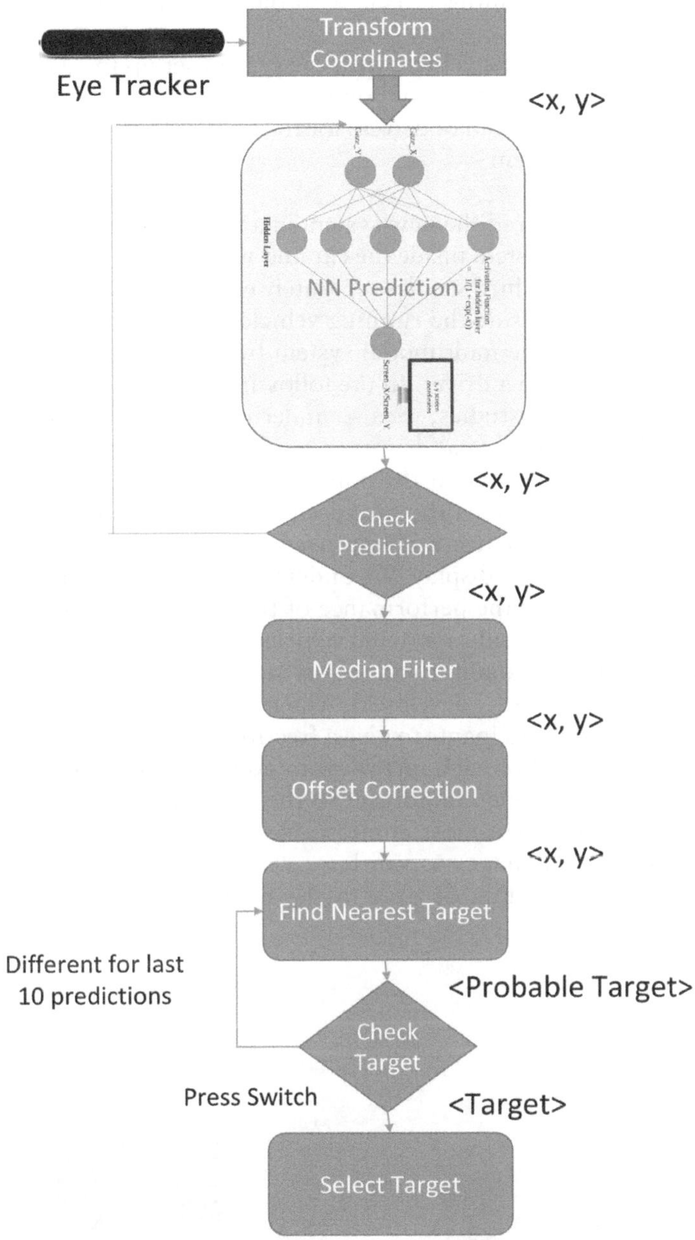

Figure 10.2 Controlling cursor in a curved plane. The flowchart explains the intermediate steps of converting eye gaze coordinates on a vertical plane to the curved surface of a windscreen using a feedforward backpropagation artificial network and then activating the target nearest to the gaze point on the windscreen.

3. Study inside a running vehicle with the finger tracking system set up for passengers
4. Study inside a static vehicle with the system set up for the driver
5. Study involving a simulator comparing physiological parameters estimating cognitive load of drivers for the final prototype and an existing touchscreen system

The first study compared the new system in a dual task setting. The second study evaluated the system inside the car and with respect to onboard vibration. The second and third studies separately evaluated the eye gaze and finger tracking system inside the running vehicle. The fourth study evaluated the performance of the multimodal system by positioning the sensors and display with respect to a driver. In the following sections, we describe these studies. After the user studies, we also undertook qualitative analysis of the system involving professional drivers.

Our user studies found that the pointing and selection times are not significantly different between the proposed system and touchscreen display; rather, driving performance was improved with the proposed system than the existing touchscreen display. We undertook a series of studies inside the vehicle and noted that the performance of the system was not significantly slower for vibrations inside a moving vehicle (Figure 10.3). The user studies were followed up by a qualitative analysis on user acceptance of the system.

An interview guide was developed, which consisted of open-ended questions that allowed participants to give a free-form answer. Participants were given time at the end of each interview to add different topics beyond the questionnaire that seemed important to them. Participants were asked the language that they were most comfortable with and proceeded with the interview in that language. A translator was appointed to prevent loss of information and meaning of data in the interview. Follow-up questions

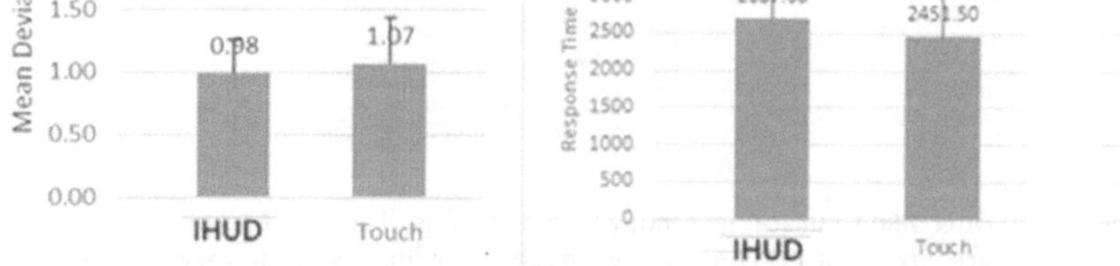

Figure 10.3 Comparing driving performance and response times of icon selection between bespoke IHUD and standard touchscreen.

were asked promptly by the researcher. After completion of an interview, the interview was translated and transcribed in English. The transcript was read thoroughly multiple times before undertaking the analysis. It helped to understand the intricate details of the life and experiences of participants and truly understand their words and actions. Interviews were conducted for over three weeks. Interviews were held for 15 to 50 minutes, with an average interview time of 20 minutes for head down display as well as HUD setup. The grounded theory method (Corbin & Strauss 2014) of analysis was used for a systematic, comparative, inductive and interactive approach to analysis. Systematic comparisons and inquiry of concepts were integrated throughout the data collection and analysis. The data were analyzed simultaneously with data collection to ensure further refined categories of data. All theoretical interpretations were explored until the most credible and reliable theory was constructed from the data transcribed from interviews. Figure 10.4 represents the theory of acceptance of gaze-controlled HUD that surfaced from saturated categories. Details of the study can be found in Prabhakar et al. (2020).

Later on, our research also developed a bespoke webcam-based eye tracker, a dataset for developing an eye tracker for automotive applications and a machine learning model for cognitive load estimation. However, within this short span, the case study aimed to demonstrate an intelligent user interface development and evaluation process involving both controlled experiments and qualitative analysis.

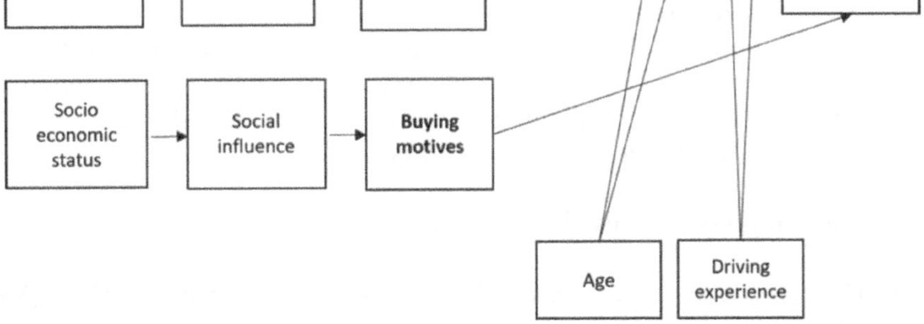

Figure 10.4 User acceptance model of gaze-controlled HUD. The figure shows a thematic network derived from thematic analysis of semistructured interviews of professional drivers about the use of the interactive HUD in a driving simulator.

10.3 GAZE-CONTROLLED INTERFACE FOR AVIATION

In the previous section, we briefly discussed an eye gaze and finger movement-controlled interactive HUD for cars with automatic distraction and cognitive load estimation features. We did not elaborate much on the development part but emphasized both quantitative and qualitative evaluation. In this case study, we will use the same effort for a more challenging domain – developing gaze-controlled displays for military fast jets. In the following section, we will briefly present the chronology of development without diving into too many details of individual parts. This section aims to give a brief outline of the development of an intelligent interaction design from simulation study to integration challenge.

In present days, a fighter aircraft pilot often undertakes a plethora of mission-related tasks in addition to the primary flying task in adverse conditions, which substantially increases the pilot's cognitive load. Pilots' cognitive load can vary with the nature of the task undertaken, environmental effects like turbulence, visibility, aircraft handling qualities, the pilot's mental state, training level of the pilot and so on. The dependability and relevance of pilot opinion ratings are always debatable. Accurate and consistent data that quantify the handling qualities of a specific aircraft are difficult to acquire. Cognitive load can be estimated by measuring multiple physiological parameters like ocular parameters, EEG and breathing rate. It may be noted that cognitive load is not measured, but rather estimated through a set of physiological parameters. These parameters are correlated with ground truth metrics or compared across experimental conditions. Cognitive load can be one of many techniques to compare design alternatives. Unlike NASA TLX or IBM SUS, there is no fixed value of cognitive load, which can be termed as good or bad. Our research considered the following ocular parameters:

- It can be measured non-invasively using commercial-off-the-shelf (COTS) sensors
- It can be used with a flying helmet and with both a clear or dark visor
- Existing eye tracking sensors can be used in a wide range of environmental conditions with high accuracy (0.4° of visual angle of accuracy) and sampling frequency (100 Hz)

Our research followed this chronology:

- Developing eye gaze-controlled desktop and virtual reality (VR) simulators
- Collecting data in actual flight using a commercial eye gaze tracker
- Investigating the relationship between ocular and flight parameters

- Undertaking simulation studies on gaze-controlled interfaces and cognitive load estimation in desktop and VR simulators
- Developing a bespoke eye gaze tracking algorithm

In the remaining part of the section, we briefly describe each step. Further details can be found in the listed research papers.

We designed the flight simulator to conduct studies in dual-task settings. Using our setup, participants undertook standard flying tasks in parallel to representative pointing and selection tasks. This setup allowed us to measure not only pointing and selection times but also the total response time consisting of the time required to switch between primary flying and secondary mission control tasks. We set up the flight simulator with both a head-down and an interactive head-up display. We also used a VR headset with an in-built eye gaze tracker to undertake studies in VR environments (Figure 10.5). We developed bespoke programs to operate both head-down and head-up display using eye gaze, similar to the previous case study described in Section 10.1. Our studies found that the gaze-controlled interface reduced response time for bigger peripheral buttons of a multifunction display (MFD), while smaller buttons were still quicker to select using a touchscreen. The use of gaze-controlled interface did not degrade flying performance; rather, it was best in one of the adaptive conditions, and users' subjective ratings about cognitive load and preference also favoured

Figure 10.5 Desktop and VR flight simulators. It does not depend on the software whether the system is AR/VR or desktop based, rather it depends on rendering media. Both computers are running the same software; the left one is used as a desktop flight simulator. If the operator moves his head, he cannot see the interior of the cockpit. The right-hand system is configured as a VR system; if the operator moves her head, she gets an immersive feeling inside the cockpit.

the gaze-controlled interface over touchscreen. The overall response time was not different across the three conditions being lowest for the adaptive condition.

We collected data from actual flights to test whether ocular parameters can be collected and correlated with flight parameters under high G-load and bright illumination conditions. Our study found that the fixation rate statistically significantly correlates with the descent rate of the aircraft (Table 10.1 and Figures 10.6 and 10.7). Details of the study can be found in Babu et al. (2019).

Finally, we developed a bespoke eye tracking helmet using data collected in actual flight conditions. It was observed that external illumination and the COTS eye tracking device's limited tracking field of view (FoV) caused ~40%–50% of gaze data loss in both flights. The proposed model received greyscale images of the left and right eye with a resolution of 192×192 and predicted normalized 2D gaze points. An attention branch was added in parallel to the feature-extraction branch to perform this intended weighted feature manipulation. Both the feature-extraction branch and attention branch contained shared convolutional layers that take eye images as the input. The feature-extraction branch produces feature vectors from both left and right eye images, while the attention branch provides the necessary weight vectors for both left and right eye features. The sigmoid activation function was used for the last layer of the

Table 10.1 Flight profiles used to collect ocular data

Flight number	*Objective*	*Profile*
Flight #1	Manoeuvring flight with head-mounted eye tracker on observer pilot	Take-off – climb – level flight to local flying area – constant G (3G and 5G) level turns both sides each – vertical loop – barrel roll – air-to-ground dive attack training missions – descent – ILS approach and landing
Flight #2	Manoeuvring flight with head-mounted eye tracker on pilot in command	Take-off – climb – level flight to local flying area – constant G (3G and 5G) level turns both sides each – vertical loop – barrel roll – air-to-ground dive attack training missions – descent – ILS approach and landing
Flight #3	Non-manoeuvring flight with head-mounted eye tracker on pilot in command	Take-off – climb – level flight to local flying area – straight and level cruise with gentle level turns – descent – ILS approach and landing
Flight #4	Manoeuvring flight with head-mounted eye tracker on pilot in command	Take-off – climb – level flight to local flying area – air-to-ground dive attack training missions – descent – ILS approach and landing

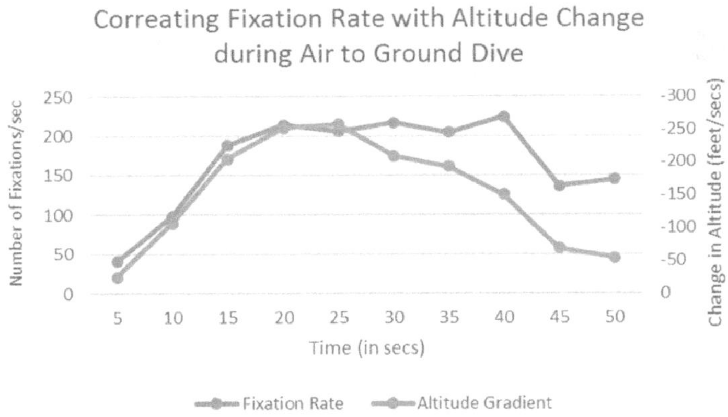

Figure 10.6 Fixation rate varies according to the rate of descent. Summarizing data over seven air-to-ground dive attacks in a Hawk Trainer aircraft, we reported a statistically significant correlation between the rate of descent of the aircraft and the fixation rate of the pilot.

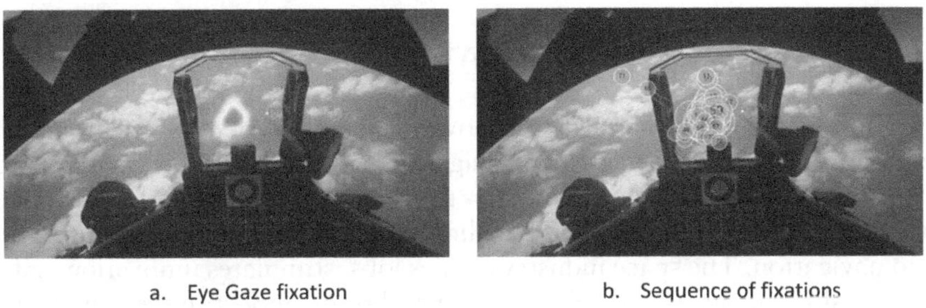

Figure 10.7 Visualization of ocular parameters during air-to-ground dive attacks.

attention branch. The feature vectors obtained from both eyes are multiplied with the corresponding weight vectors to obtain weighted features that will be passed further through the network for the regression task. A global average pooling (GAP) layer was added after convolutional layers to obtain single-dimensional feature vectors from feature maps. This step is computationally more efficient than flattening the feature maps from convolutional layers. The ReLU activation function was used for all convolution, dilated convolution and fully connected layers in the network, except the last layer of the attention branch and model output layer, where sigmoid was used. The whole model was trained with a laboratory-collected custom dataset with subject and illumination variations. When tested on actual flight data, the proposed system was found to be more consistently tracking eye gaze in high illumination (>3000 lux) conditions than COTS devices (Figure 10.8).

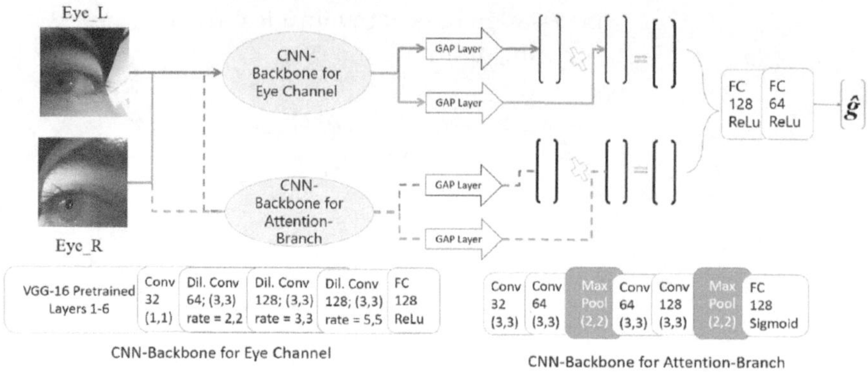

Figure 10.8 Network architecture for helmet-mounted IR gaze estimation.

In the final section of the chapter, we will go from aviation to spacecraft design and present VR simulator design and evaluation for a de-orbiting task.

10.4 VR SPACECRAFT SIMULATOR

Our foray into space has led to innovations with varied practical applications on Earth such as satellite navigation, advancements in material sciences, robotics and medical research to name a few. National security has become increasingly reliant on satellite-based surveillance, communication and navigation. The space industry creates jobs; stimulates innovation, entrepreneurship, international collaboration and trade opportunities; and fosters the next generation of scientists, engineers and explorers. Space technology stokes our curiosity to explore. The first human-crewed spacecraft orbited Earth in 1961 (within four years of the launch of the first artificial satellite) and man first stepped on the moon in 1969. Since then, space stations such as Salyut, Mir, Skylab, ISS (International Space Station) and Tiangong have built our understanding of sustaining long-duration stays in space. For the purpose of this research, we define manned transport spacecraft as those vehicles that ferry humans from Earth to space and back and are capable of autonomous orbital flight. All practically useful spacecraft requires the ability to control their flight path. Flight path control implies the ability to determine and change position and orientation (attitude). Typical flight path control tasks of manned transport spacecraft are as follows:

- Achievement and maintenance of solar power generating attitude – Pointing the electrical power generating side of the spacecraft's solar array towards the sun

- Orbit raising/circularization – Orienting the spacecraft to point the main engine thrust vector along the velocity vector and fire the engines at a predetermined time/orbital position to increase orbital velocity by a specific amount (delta-V)
- Rendezvous and docking
- De-orbit – Orienting the spacecraft to point the main engines thrust vector opposite to the velocity vector and fire the engines at a predetermined time/orbital position to decrease orbital velocity by a specific amount (delta-V). This puts the spacecraft into a trajectory that re-enters the Earth's atmosphere
- Atmospheric flight path control (distinct from the on-orbit tasks)

On-orbit manual attitude control of manned spacecraft is accomplished using external visual references and some method of three-axis attitude control. All past, present and developmental spacecraft feature the capability to manually control attitude for de-orbit. National Aeronautics and Space Administration (NASA) spacecraft permit an aircraft windshield type 'front view', wherein an arc of the Earth's horizon is visible to the crew in de-orbit attitude. Russian and Chinese spacecraft permit the crew a 'bottom view', wherein the entire circular Earth horizon disk is visible to the crew in de-orbit attitude. Our study compared these two types of external views for efficiency in the achievement of de-orbit attitude. We used a bespoke Unity Virtual Reality (VR) spacecraft simulator. The task was to accurately achieve de-orbit attitude while in a 400 km circular orbit. Six military test pilots and six civilians with gaming experience flew the task using two methods of visual reference. Comparison was based on time taken, fuel consumed, cognitive workload assessment and user preference. We used ocular parameters, EEG, NASA TLX and IBM SUS to quantify our results. Our study found that the bottom view was easier to operate for the manual de-orbit task (Figure 10.9). In a second study, we compared

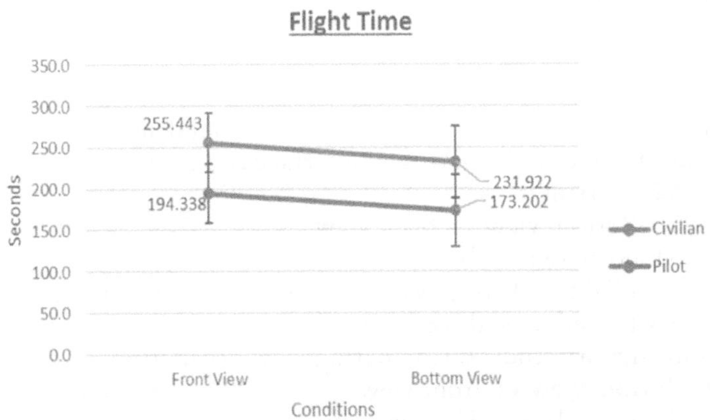

Figure 10.9 **Task completion time versus view for different participants.T**

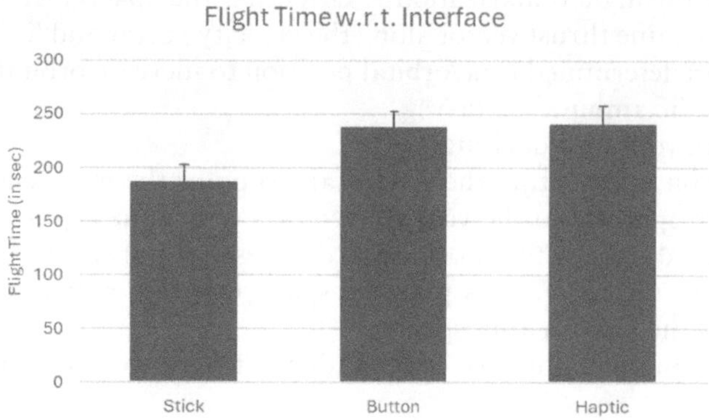

Figure 10.10 Task completion time versus interface.

three different devices to control attitude: a 3-DoF flight stick, physical buttons and virtual buttons with haptic feedback. While the flight stick was fastest in terms of de-orbiting, physical buttons were most economical in terms of fuel consumption (Figure 10.10). Additionally, we realized that a VR based system can work as a training simulator for manual on-orbit flight path control tasks by astronauts (both pilots and non-pilots).

Task description

- Initial attitude error with respect to de-orbit attitude:
  - Pitch 0°
  - Roll 102°
  - Yaw 104°
- A portion of Earth was visible in either of the cameras.
- Task completion criteria:
  - Pitch ±1°
  - Roll ±1°
  - Yaw ±6°

Three user studies

- Pilot study: VR vs. 2D interface involving students
- Study 1: View and manoeuvre interface comparison
  - View orientation
    - Bottom view vs. front view
  - Manoeuvre interface
    - Flight stick vs. physical buttons vs. virtual buttons
  - Six test pilots and six civilians
- Confirmatory study 2: Comparing view orientation
  - Bottom view vs. front view
  - Six test pilots and six civilians

List of instruments

- HTC Vive Pro Eye – VR headset with integrated Tobii Eye Tracker
- Emotiv 36 channel EEG tracker
- Emotiv EEG Data Analysis Suite
- Thrustmaster 3 DoF Flight Stick
- Manus Haptic Gloves
- Cambridge EDC Inclusive Design Toolkit for space suite simulation
- Bespoke VR software
- Bespoke physical switch

The proposed study demonstrated that a VR-based system can work as an excellent training simulator where both test pilots and civilians could learn de-orbiting manoeuvres and undertake the same in less than 5 minutes with realistic control laws. Either view could be used to de-orbit effectively. Heading correction in front view required holding the inertial attitude steady and observing roll error build-up. No such hold and observe was required for the bottom view. This explains why the front view resulted in longer task achievement time than the bottom view. In terms of fuel consumption, the front view resulted in lesser fuel consumption for test pilots than the bottom view. In near local vertical attitude, small errors in pitch/bank showed up more prominently in the bottom view than the front view. Pilots tended to correct these errors while fine-tuning heading, resulting in higher fuel consumption in the bottom view. Participants reported less cognitive load through NASA TLX and higher preference through the IBM SUS questionnaire for the bottom view compared to the front view. This may be explained by the additional effort required for close observation to determine heading error in the front view, indicative of lower cognitive workload with bottom view.

Pilots with prior spacecraft training reported that the task and simulation were realistic. One pilot felt that the display flicker with the VR headset caused eyestrain, which was not representative of actual spaceflight. However, this flicker was common to both views and hence did not affect this study. It was realized that the front view required accurate achievement of near-zero bank with respect to local vertical to correctly appreciate heading error. In the case of any bank while correcting heading (e.g., inadvertent pilot input, control cross-coupling and perturbing torques), it would be very difficult to proceed with further heading correction in the front view for the de-orbit task. While this was a drawback, the front view also provided a design advantage wherein there was no requirement of a zoom camera for heading correction. The front view also had obvious advantages for docking as docking ports tend to be front aligned.

The initial attitude error determined through the position of the Earth on the display resulted in nearly all participants following a similar control strategy. The sequence followed was coarse roll correction followed by fine pitch and roll correction. Once roll and pitch were approximately correct, participants applied coarse yaw correction followed by fine yaw correction. Participants gave fine touch-ups to all three axes of control before signalling completion of the task. This was true for both viewing methods.

Figure 10.11 shows two representative heatmaps of eye gaze fixations comparing top and bottom views. The eye gaze fixations are generated from the whole duration of the task and superimposed on the final image. The front view shows eye gaze fixations on the upper part of the image based on the initial position of the Earth. The bottom view, rather, shows the Earth's position always in the middle of the view, and the heatmaps are concentrated in the middle of the screen. The heatmaps for both bottom and front views show that eye gaze tended to fixate at random and had saccadic movements between the top and bottom during coarse pitch and roll corrections. Coarse yaw correction in bottom view also followed the same trend. During fine pitch and roll corrections, participants tended to fixate on the horizon and reference marks drawn on the user interface to match them correctly. Fine yaw correction in the bottom view required the participant to view minute terrain or cloud features along drift markers, which are evident on the bottom view heat maps. Coarse as well as fine yaw correction in the front view required careful observation of the horizon marks

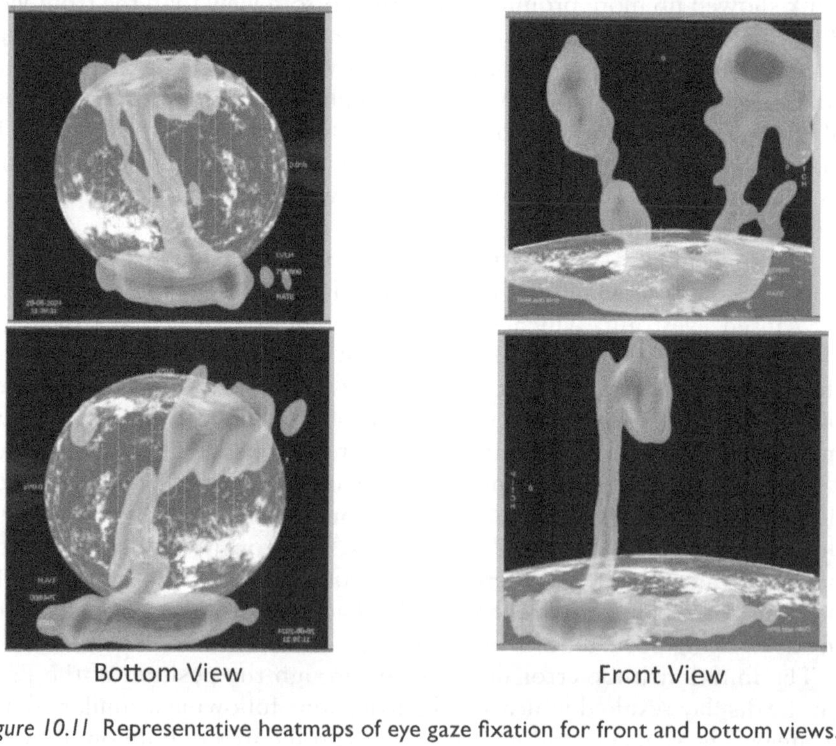

Figure 10.11 Representative heatmaps of eye gaze fixation for front and bottom views.

against the earth horizon; this was evident in the heat map. The heat maps faithfully replicated the control strategy followed by participants.

Control stick and physical buttons were both effective to achieve the task. All participants reported that the virtual buttons did not accurately mimic the touchscreen primarily due to a lack of depth perception. The control stick was preferred to the buttons. This can be explained by the fact that the stick was spring-loaded to centre position and hence gave direct haptic force feedback of angular rate. Spacecraft rotation could be stopped simply by releasing the stick. Participants did not have to refer to the numeric angular rate demand values on the display while using the control stick. However, effective button control could only be done using the numeric angular rate demand values on the display. A surprising effect of using the stick was that since participants were more confident in applying minor corrections with the stick, it resulted in higher fuel consumption. Details of the study can be found in Krishnan et al. (2024, 2025).

In summary, this study presents a comparison of viewport orientations for de-orbiting manoeuvres of a spacecraft using a bespoke virtual reality simulator. Based on data from six test pilots and an equal number of civilians, the bottom viewport is found to be easier to operate than the front viewport for the de-orbiting manoeuvre. Besides the immediate result, the study also demonstrates the use of a VR simulator and a set of human factor analysis tools for the investigation of the human–machine interface of spacecraft.

10.5 QUICK FACTS

- **HUD**: A head-up display is a see-through display in the cockpit of an aircraft/car that projects essential information on the windshield directly in a pilot/driver's forward field of view. This reduces the need to look inside the cockpit
- **Glass cockpit**: A glass cockpit consists of multiple electronic MFDs that replace traditional flight instruments. It simplifies aircraft operation and navigation and allows pilots to focus only on the most pertinent information
- **MFDs**: Multifunction displays are part of the glass cockpit. Unlike showing all information together like analog gauges, MFDs allow only relevant information to be shown and changed. They are operated by touchscreen or bezel keys on the perimeter
- **HMDS**: The head (or helmet)-mounted display system projects essential flight and targeting information directly on the visor of a helmet, thereby enabling full head movement without the need to look forward for flight information. It enhances the field of view and weapon aiming ability as compared to a HUD

- **ADI**: The attitude direction indicator of an aircraft or spacecraft provides a reference of the vehicle with respect to the horizon. It mainly shows an attitude ball with brown and blue colours indicating ground and sky, respectively. For spacecraft, the contour of the Earth is used as the horizon marking

10.6 SUMMARY

In summary, this chapter aims to consolidate our discussion on designing and evaluating intelligent user interfaces. Although we addressed three different application areas, the similarities in the approach should be noted in terms of design, development and integration with simulators, sampling of participants and user study design. While the case studies investigated the integration of a specific technology, a cockpit design is different from traditional consumer electronics interface design in terms of strict adherence to existing international and military standards. In Chapter 12, we will investigate related international standards and standard development organizations for general user interface and cockpit design.

Chapter 11

Case studies on trajectory prediction

11.1 INTRODUCTION

The previous chapter discussed cockpit design and integrating novel modalities of interaction. Pointing and selecting targets in displays in different configurations (head down, head up, head mounted) can be challenging. In this chapter, we will investigate technology to facilitate human–machine interaction for direct manipulation interfaces, which allow for directly pointing and selecting targets in a graphical user interface (GUI). Target prediction or intent recognition is well investigated in the field of human–machine interaction for reducing pointing times while operating GUIs for both physical and situational impairment. Earlier work on target prediction worked on 2D coordinates, but later research for touchscreen systems or automotive dashboards investigated 3D coordinates. Situational or physical impairment introduced new challenges in target prediction as the trajectory turns complex and non-linear. However, many previous works did not investigate how the prediction model can be ported for a new task or situation.

This chapter will discuss four different applications for trajectory prediction or intent recognition. The first use case will utilize the concept of Fitts' law discussed in Chapter 2. We proposed an artificial neural network-based model to classify the homing phase from the ballistic phase and predicted the target once we detect that the user is trying to home in on a target using multiple modalities of interaction. The second and third use cases explored imitation learning, so that the prediction model can be easily retrained with sample trajectories for any new situation. We will discuss trajectory prediction for XR interfaces and human–robot handover requests. The last use case will compare signal processing and machine learning-based approaches for walking trajectory prediction for a VR digital twin of office spaces.

11.2 HAND TRAJECTORY PREDICTION

Target prediction algorithms are beneficial for people who find a regular pointing device difficult to use. Two possible scenarios are users with motor

DOI: 10.1201/9781003646716-11

impairment and first-time users of a novel interaction device. There is a plethora of studies demonstrating the difficulty of people with age-related or physical impairments in stopping pointer movements or clicking on small targets. Another possible scenario is using a novel interaction device involving head or eye-gaze movement. Devices like head or eye trackers are not as easily controllable as a desktop mouse or touchpad, and people may struggle to use these devices to precisely control pointing movements. Devices like head or eye trackers have many applications for both physical and situational impairments (e.g., operating assistive software by disabled users using eye trackers and locking a target by fighter aircraft pilots using head trackers). We have implemented a bank of neural networks considering all possible combinations of three movement properties (bearing, velocity and acceleration of movement). We have used supervised learning to detect users' movement phases from movement features. Initially, we trained the neural networks with a multiple distractor task. Later we used the same task to evaluate the networks following our evaluation criteria using different training and test sets of participants.

We conducted three user trials involving three different input modalities:

1. The first trial involved people with age-related and physical impairments, and they used a standard computer mouse
2. The second trial involved a head tracker and brain–computer interface used by able-bodied users who never used a similar system before
3. The third trial involved an eye-gaze tracker used by able-bodied users who never used a similar system before

The study was conducted using a 21-inch screen (435 mm × 325 mm) with 1600 × 1200 pixels resolution and a standard computer mouse. The head tracking was conducted using an Emotiv Epoc brain–computer interface headset, while a Tobii X120 eye tracker was used for eye tracking. We have defined the following three parameters to evaluate the quality of a target prediction algorithm:

- **Availability**: In how many pointing tasks the algorithm makes a successful prediction
- **Accuracy**: Percentage of correct predictions among all predictions
- **Sensitivity**: How quickly an algorithm can detect the intended target

Figure 11.1 plots the average percentage of availability and accuracy in different systems. In the figure, different combinations of features are plotted on the x-axis, while the average availability and accuracy are plotted on the y-axis. The model only fires (or turns available) when it detects a change in movement phase. In the availability graphs, green bars show the percentage when it correctly identified the target, the red bars show when it could not.

On certain occasions, it fails to detect this change of movement phases, and so the availability is not always 100% (white bars). Figure 11.1 also plots the sensitivity of the system. The x-axis shows the fraction of pointing time spent on a scale of 100. The y-axis represents the probability of correct target prediction. We found that velocity and bearing of movements have the highest availability for mouse and eye-gaze tracking systems, while the combination of all parameters has the highest availability for the head tracking system. Velocity and bearing also have the highest accuracy for all systems. The sensitivity is found to be rather the same for all possible combinations of parameters. Video demonstrations of the multiple distractor task are available at http://youtu.be/QsxDxcccwAw, and the target prediction system is available at http://youtu.be/p9YOKj59TiY. The proposed model accurately detects changes in the pointing phase in more than 70% of the pointing tasks for head and eye-gaze trackers and 65% for mouse users with age-related or physical impairments. The accuracy of target prediction is nearly 60% for all cases; however, it reaches more than 90% as the user approaches the target, as shown in the sensitivity graph (Figure 11.1). We have developed an adaptation system that enlarges a target whenever it is predicted as a probable target; whenever the target prediction system predicts a target, we increase its size by 1.5 times. We do not change the colour or any other property of the target. A user study involving six able-bodied people using a head tracker and 181 pointing tasks found that the target prediction system can reduce pointing time when coupled with the target magnification system. Details of the study can be found in Biswas and Langdon (2014).

11.3 TRAJECTORY PREDICTION FOR XR INTERFACE

In extended reality (XR) environments like virtual reality (VR) and mixed reality (MR), users perform pointing tasks to select targets. Accurate target prediction helps improve interaction speed and user experience. However, XR systems often lack haptic feedback, making early predictions particularly important. Traditional methods struggle to capture the complex hand dynamics and user preferences involved in these interactions. This study used inverse reinforcement learning (IRL) to predict one of many virtual buttons (later referred to as targets) in XR. Predicting the target early allows the system to respond faster. It highlights the intended target before the hand reaches it. This reduces selection time and errors. Prediction was based on partial hand trajectories. The proposed method used sampling-based maximum entropy inverse reinforcement learning (SMEIRL). SMEIRL models an agent operating within a Markov decision process (MDP), optimizing an underlying reward distribution. MDP is a mathematical framework used to model decision-making in uncertain environments. It’s a type of stochastic process where the future state of a system only depends on its current state,

Figure 11.1 Analyses of target prediction systems.

not on the past history, a property known as the Markov property. As discussed in Chapter 3, in reinforcement learning, rewards indicate a short-term benefit or loss to an agent based on its current position in state space relative to its goal state. In the present case study, SMEIRL-modelled hand movement as an agent optimizing a reward function, which is calculated based on the present hand position relative to a set of virtual target buttons. It approximated the reward by generating sample trajectories. A custom sampler created trajectories following the shortest path and a quadratic velocity profile based on hand velocity data. The algorithm predicted the target by calculating probabilities for each possible target based on partial hand movement. Figure 11.2 shows the prediction accuracy as a function of task completion percentage. In MR, SMEIRL reaches approximately 72.5% accuracy by 50% task completion, increasing to 98% accuracy by the end of the task. In VR, it achieves about 58% accuracy at 50% completion, rising to 96% at task end. Compared to classical kinematics ($CM_{k=5}$), which reached a full convergence point (FCP) only after 80%–90% of the task, SMEIRL predicts the target much earlier. Early high accuracy indicates the system's ability to anticipate user intent quickly during the rapid aiming motion.

Figure 11.3 shows the average prediction accuracy from a given point in the task until completion. SMEIRL maintains cumulative accuracy above 85% from 50% task completion onward in MR and above 78% in VR. MEDIRL (maximum entropy deep IRL) and classical kinematics showed lower cumulative accuracy in the early to mid-task range, with SMEIRL outperforming both consistently. This stability means SMEIRL maintained correct predictions once it identifies the intended target, which is critical for a seamless user experience in XR.

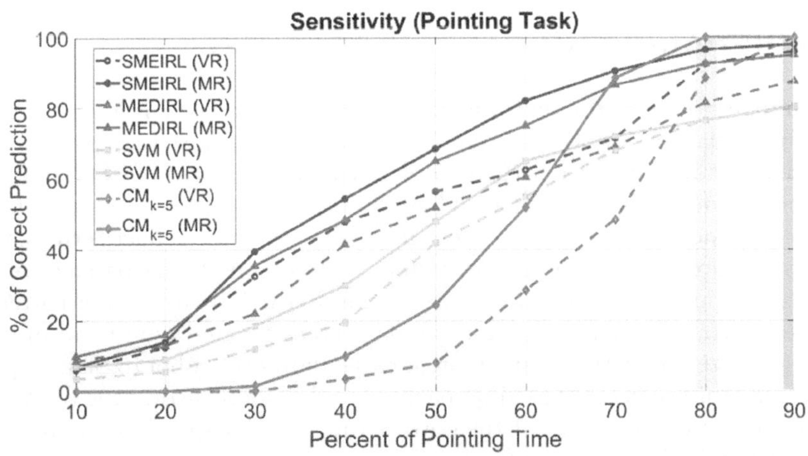

Figure 11.2 Sensitivity of target prediction for the pointing task in VR and MR. $CM_{k=5}$ reaches FCP at 80% and 90% completion in MR and VR, as shown by the grey regions, respectively,

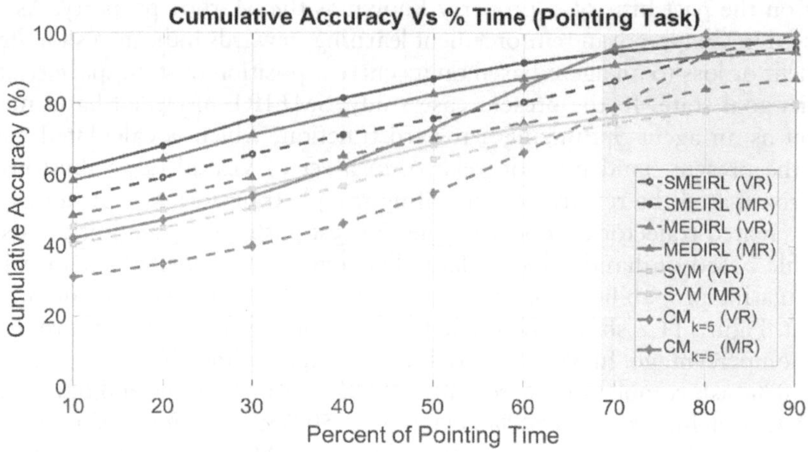

Figure 11.3 Cumulative accuracy with time for the pointing task in VR and MR. The cumulative accuracy of SMEIRL surpassed that of MEDIRL throughout the pointing task in VR and MR. The cumulative accuracy of $CM_{k=5}$ exceeded that of SMEIRL in the grey region, ranging from approximately 68% to the task's completion.

SMEIRL effectively predicted user-intended targets during the pointing task in XR. It outperformed other state-of-the-art methods in both prediction accuracy and the minimal input trajectory length required, improving the responsiveness of XR systems and addressing challenges due to the lack of haptic feedback. Details of the study can be found in Mitra et al. (2023).

11.4 HUMAN–ROBOT HANDOVER REQUEST

In human–robot interaction, intent prediction plays a key role in ensuring smooth collaboration. During handover tasks, robots must anticipate human movements and intended targets to plan safe and efficient object handover. Human preferences vary widely in handovers, including start positions, hand velocities and takeover points. Accurate intent prediction reduces errors, prevents collisions and improves ergonomics. Traditional physics-based or probabilistic models struggle with complex, non-linear human motion patterns. This study applies IRL to predict human intent by analyzing hand motion and eye gaze. Predicting human intent early allows the robot to plan its path safely. Intent prediction based on partial hand motion ensures the robot can take control of the object at the correct time and location. Combining eye gaze data with hand motion further enhances prediction accuracy by providing additional cues about the human's focus and goal. The approach uses MEDIRL. MEDIRL employs a neural network to model complex, non-linear reward functions reflecting human preferences during handovers. The model

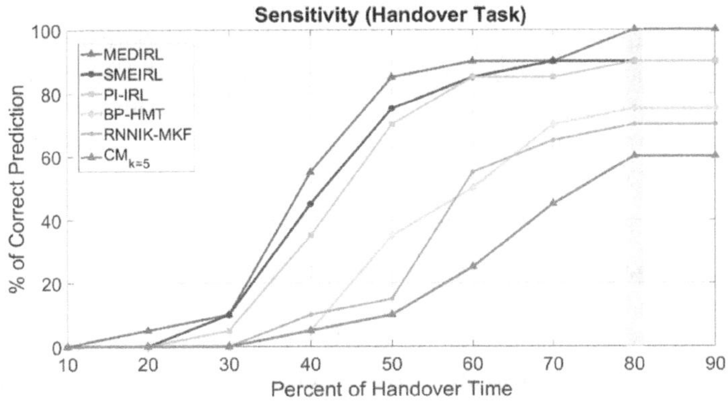

Figure 11.4 Sensitivity of target prediction during human–robot handover. MEDIRL reaches FCP at 80% task completion as depicted by the grey region.

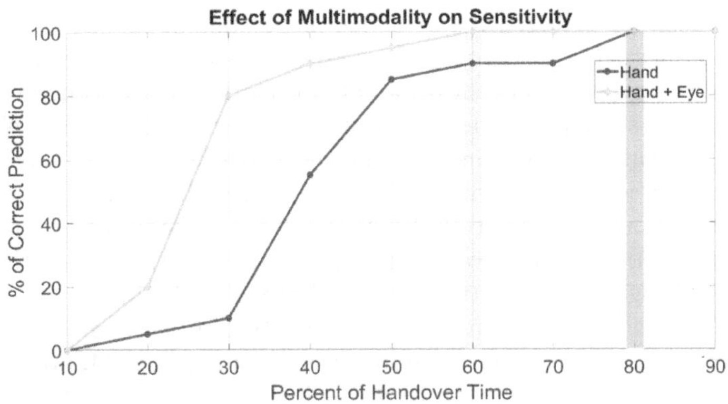

Figure 11.5 Sensitivity of target prediction using MEDIRL. The multimodal prediction method achieves FCP earlier, at 60% task completion, compared to hand-only prediction, which reaches FCP at 80% task completion.

learns from demonstrations, capturing hand velocity and movement patterns. Features include distance to target, velocity, acceleration, jerk and gaze fixation data. Eye gaze is integrated by weighting target probabilities based on average fixation duration (AFD), improving prior estimation of intended goals. Figure 11.4 shows the prediction accuracy rising quickly during the handover task. Using hand motion, MEDIRL reaches 86.4% accuracy at about 50% task completion. It reaches FCP (99.9% accuracy) at 80% task completion, the highest compared to other methods.

Adding eye gaze data, accuracy reaches 99.9% at 60% task completion, as shown in Figure 11.5. Combining eye gaze data with hand motion increases prediction accuracy and reduces prediction time, enabling faster robot response.

MEDIRL effectively predicts human intent during rapid handovers in human–robot interaction. Its deep learning framework captures complex motion dynamics, outperforming prior physics-based and probabilistic models. Integration of eye gaze data with hand motion improves both prediction accuracy and minimal input required for prediction, enabling robots to respond swiftly. Details of the study can be found in Mitra et al. (2024, 2025).

11.5 WALKING TRAJECTORY PREDICTION

Walking trajectory prediction is mostly explored in autonomous driving situations to predict pedestrian movements. We investigated walking trajectory prediction for a different use case involving a VR digital twin of office space. During 2020 to 2025, we were collaborating with British Telecom to develop a privacy preserving technology to estimate room occupancy and energy consumption in built environments. We developed and deployed a system where we detected the presence of persons in an office space using multiple cameras but did not recognize their identities for privacy purposes and then redrew avatars in a VR model of the office space. A demonstration of the overall system can be found at https://cambum.net/BT/. While we could draw avatars of persons in standing or sitting posture, generating an animation for movement was challenging. All walking animation models require a start and end position; however, in real time we could only record the instantaneous position of a person. Hence, we investigated the prediction of walking trajectory to give an end point to the animation model to generate realistic walking animation in the VR digital twin.

The proposed system used both machine learning (LSTM, GRU) and signal processing approaches (particle filter, Kalman filter). LSTM (long short-term memory) is a type of recurrent neural network that captures long-term dependencies in time-series data. GRU (gated recurrent unit) is a more lightweight variant of a recurrent neural network compared to LSTM, which is computationally more efficient while still capturing essential temporal dependencies. Particle filter (PF) is a probabilistic sequential Monte Carlo method that uses a set of weighted particles to estimate the state of a system. It does not require training data and is particularly effective for non-linear, non-Gaussian motion. The Kalman filter (KF) is a linear estimator that operates in a predict–update cycle. It smooths noisy trajectory data and provides basic trajectory estimation based on state-space modelling. Overall, the study employed YOLOv8 for real-time person detection using RGB cameras. The extracted foot positions were converted from pixel space to real-world coordinates using third-degree polynomial regression. A velocity threshold of 2 m/s determined using K-means clustering was used to trigger prediction models.

The performance of the models was evaluated using metrics such as accuracy, root mean square error (RMSE) and discrete Fréchet distance (DFD). Prediction was assessed at different percentages of trajectory completion,

50%, 70% and 90%, to determine early mid- and late-stage accuracy. Figure 11.6 shows the accuracy plot at 50% LSTM achieved 47.62% accuracy, outperforming PF (29.23%) and KF (24.62%). At 70%, LSTM improved to 85.71%, followed by PF (64.61%), GRU (76.19%) and KF (43.07%). At 90%, LSTM (90.48%) and PF (89.23%) performed comparably, demonstrating PF's ability to catch up despite not relying on training data. PF achieved the lowest RMSE (0.41 m), outperforming KF (0.80 m), LSTM (0.91 m) and GRU (1.08 m).

The DFD is a measure of similarity between two polygonal curves, also known as trajectories or sequences of points. In terms of DFD, as shown in Figure 11.7, which measures trajectory shape similarity, PF again led with

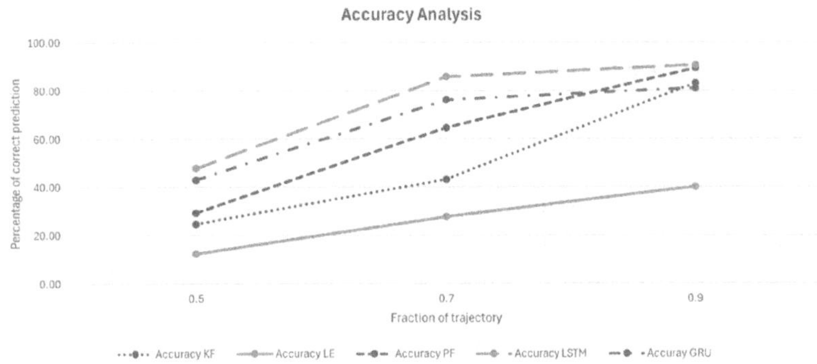

Figure 11.6 Accuracy plot of various prediction models.

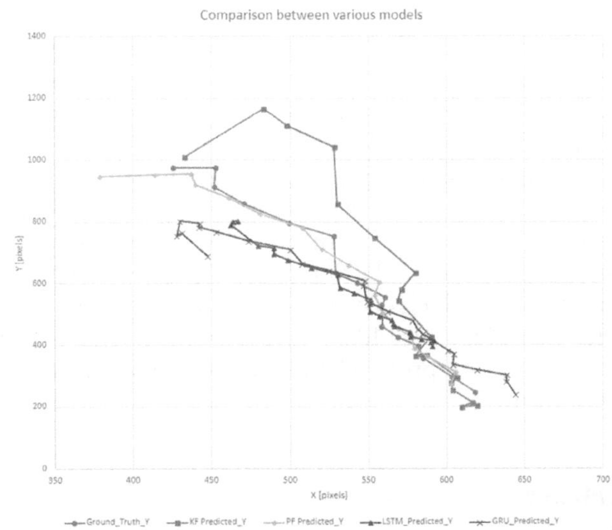

Figure 11.7 DFD analysis for methods with respect to ground truth trajectory.

a value of 54.20, versus KF's 192.35, LSTM's 221.45 and GRU's 284.97. This makes PF particularly valuable for applications where the path shape matters more than the final position alone.

This work demonstrates that LSTM excels in accuracy (90.48%) given sufficient training data, while particle filters provide superior robustness (71.88% better DFD) for dynamic environments. Kalman filter and GRU serve as lightweight options but fall short in handling complex, non-linear motion patterns. By integrating these models into a VR-based digital twin, the study demonstrates how prediction improves user experience in immersive environments. Details of the study can be found in Talwar et al. (2025).

11.6 QUICK FACTS

- **Trajectory prediction**: A set of algorithms that can predict or extrapolate a partial trajectory
- **Target prediction**: A set of algorithms that predict one or more possible targets from a set of possible targets. Unlike trajectory prediction, the algorithm does not predict trajectory but calculates the most likely target at the end of a trajectory
- **Imitation learning**: A machine learning technique where an agent learns from demonstrations
- **Inverse reinforcement learning**: An algorithm where we calculate a reward distribution over states (may be considered positions for trajectory prediction) from demonstrations (sample trajectories) and then use a standard reinforcement learning algorithm to learn policy (trajectory)
- **Particle filter**: Particle filter, or the sequential Monte Carlo method, is an algorithm where we estimate the posterior probability distribution of future possible states of a stochastic process using a set of sample probability distributions known as particles. The weights of individual particles or samples are updated at each step based on prior probability or present measurement. It was used to predict trajectories by simulating many possible future positions based on current velocity and motion uncertainty. Each particle represented a possible state of a human detected, which evolves through the motion model. Over time, particles that align with likely measurements were retained and projected, producing a probabilistic estimate of where the person might go in the future

11.7 SUMMARY

Target prediction or intent recognition has been studied for many applications including aviation, social computing and human–machine interaction.

At its simplest form, target prediction may take the form of linear extrapolation to more complex activity of intent recognition or behaviour forecasting in the domain of social computing. The chapter described different algorithms in four different use cases. We started with a simple artificial neural network-based system that detects the homing phase and enlarges the nearest target similar to the flowchart (explained in Chapter 10, Figure 10.2). The second and third use cases used the concept of imitation learning, where the prediction system learns to predict trajectory from a limited number of examples. The last use case compared signal processing and machine learning algorithms for walking trajectory prediction, which also involves a computer vision module to detect persons and linear regression-based mapping between real and virtual coordinates. Together the chapter presents a summary of using AI techniques to improve the usability of multimodal direct manipulation interfaces.

Chapter 12

Standardization

12.1 INTRODUCTION

This chapter will discuss international standards. It is often said that a standard has a longer lifespan than a specific technology. For example, the International Telecommunication Union states that 'Whenever you are making a telephone call wherever in world, you are using an ITU standard'. While a complete and exhaustive listing of standards related to AI/ML and user interface design could be a book in its own right, in this chapter we will cherry-pick a few examples of standards that were mentioned or discussed in previous chapters. We will not be confined to international standards; we will also briefly discuss military standards on interface design, as those are often more detailed than corresponding international standards. The standard development process itself is often time-consuming and laborious but also rewarding in terms of understanding diversity in culture, convention and linguistics across nations. It is hoped that at the end of this chapter, readers will get an idea of the importance of standards, an overview of standard development organizations, and their applications in intelligent interface and interaction design.

12.2 WHY STANDARDS

In previous chapters we learned about technology to develop and evaluate intelligent user interfaces. Then why are we studying standards? Keeping in mind creativity, both AI and UI/UX design often need to adhere to or follow certain requirements, guidelines, or recommendations in terms of both design and evaluation. For example, we have already come across the following standards in previous chapters:

- ISO 9241 Ergonomics of human–system interaction, in the context of Fitts' law (Chapter 2)

DOI: 10.1201/9781003646716-12

- ISO/IEC 18039:2019 Information technology — Computer graphics, image processing and environmental data representation — Mixed and augmented reality (MAR) reference model (Chapter 4)
- W3C Accessibility Guidelines (WCAG), in the context of usability evaluation (Chapter 8)
- ISO 10218 Robotics – Safety requirements, in the context of describing a safe operating environment for cobots (Chapter 9)

This chapter will cover a few more standards in the context of developing and evaluating intelligent user interfaces, after learning the standard development process.

12.3 STANDARD DEVELOPMENT PROCESS

Standardization is a process of standard making, whereby a standard is a set of technical specifications adhered to by a producer, either tacitly or as a result of a formal agreement. The standard development process often starts with gathering a set of use cases and eliciting the gap in present status and future goal (Figure 12.1). A technical standard aims to develop a set of specifications, recommendations or requirements for product or service development.

Examples of standard development organizations (SDOs) include:

- International Standardization Organization
  - National mirror organizations

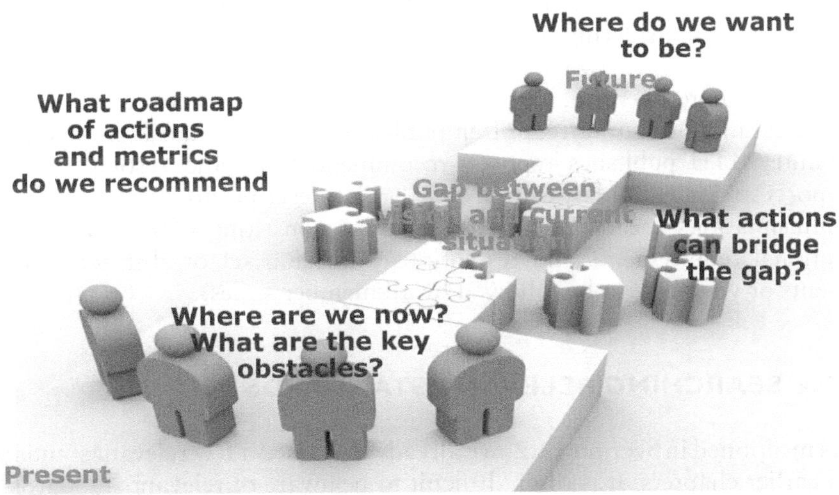

Figure 12.1 Starting point of the standard development process.

  - BIS for India
  - DIN for Germany
  - BSI for UK
  - ANSI for USA
- International Telecommunication Union
- European Conference of Postal and Telecommunications Administrations (CEPT)
- European Telecommunications Standards Institute (ETSI)

Each SDO follows its own protocols to develop standards, but there are the following similarities in the overall development process:

1. Collection of use cases – This is often published as a technical report and serves as a roadmap for the future standard development process
2. Gap analysis and initial drafting – From use case descriptions, experts try to analyze present gaps in terms of existing standards and define a set of terms, recommendations or specifications to be part of a future standard
3. Consent/agreement – Ideally, a technical standard should be followed by all member states. Each SDO has separate study committees or study groups that lead the process of standard development. The initial draft standard is discussed through a series of regular and ad hoc meetings until consent or agreement is reached from all member states. During the drafting process, other SDOs and organizations send liaison statements on similar work which are considered in terms of drafting a new standard
4. Publication and maintenance – Once consent is reached, the standard is published. Over time, a standard can be revisited, updated or discarded considering new developments

It may be noted here that while ISO publishes international standards, its national mirror committees often publish national standards for a specific country. ITU publishes reports, recommendations and resolutions. The reports are pre-standardization works like descriptions of use cases recommendations are technical specifications consisting of mandatory and optional requirements and resolutions are like a set of abstract requirements or wish lists to be followed by all member states.

12.4 SEARCHING RELEVANT STANDARDS

As mentioned in Section 12.2, we already discussed a few relevant standards in earlier chapters. It is often difficult to be aware of relevant standards or guidelines considering the sheer volume of them; however, most SDOs offer

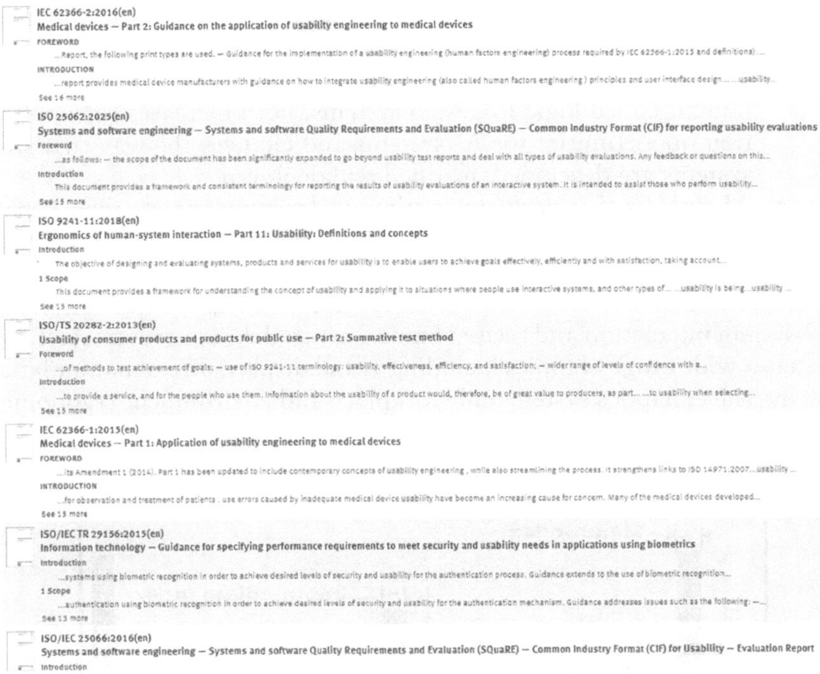

Figure 12.2 Search results for the term 'usability' in ISO database.

services where one can input a technical word and find its definition from existing standards (e.g., the ISO Online Browsing Platform at https://www.iso.org/obp/ui or the ITU Terms and Definitions database). For example, searching the ISO database with the word 'usability' on 13 May 2025 produced the results in Figue 12.2. Based on the application of 'usability', one may search the relevant standard.

12.5 EXAMPLES OF STANDARDS AND GUIDELINES

International standards often provide guidelines for both AI and UI/UX design and evaluation. However, many industries and regulatory bodies publish application-specific guidelines. In this section, we will review a few relevant standards and guidelines.

- ISO/IEC 42001:2023(en) intends to help organizations responsibly perform their role with respect to AI systems (e.g., to use, develop, monitor, or provide products or services that utilize AI). AI potentially raises specific considerations such as:

- The use of AI for automatic decision-making, sometimes in a non-transparent and non-explainable way, can require specific management beyond the management of classical IT systems
- The use of data analysis, insight and machine learning, rather than human-coded logic to design systems, both increases the application opportunities for AI systems and changes the way that such systems are developed, justified and deployed
- AI systems that perform continuous learning change their behaviour during use. They require special consideration to ensure their responsible use continues with changing behaviour

- ISO 9241 (2011) is a multipart standard about ergonomics of human–system interaction and related human-centred design processes. It covers a wide range of topics including visual display, direct manipulation interface, haptics system and workplace and environment ergonomics

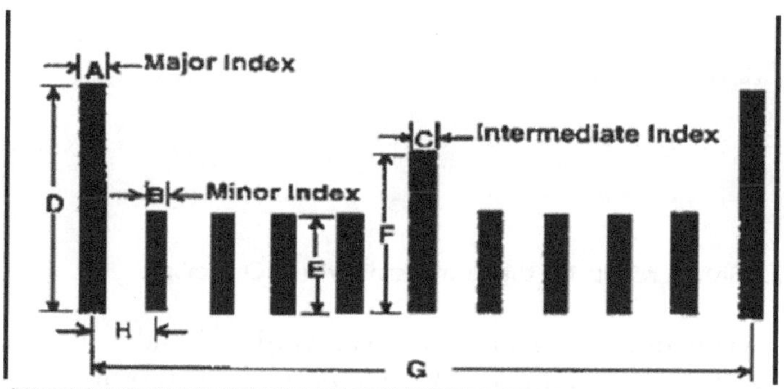

Dimensions of dark markers on light background, visual angle [1]	
A Width of major scale index	1.16 mrad (4 min) [2]
B Width of minor scale index	0.87 mrad (3 min) [2]
C Width of intermediate scale index	1.16 mrad (4 min) [2]
D Length of major scale index	7.86 mrad (27 min)
E Length of minor scale index	3.49 mrad (12 min)
F Length of intermediate scale index	5.82 mrad (20 min)
G Width of gap between major scale index	25.02 mrad (86 min)
H Width of gap between minor scale index	2.62 mrad (9 min)

NOTES:

[1] For most applications with a dark graduation mark on a light background, the width of the minor graduation mark can be used for major and intermediate graduation marks as well. Use of this strategy allows the width of the pointer tip to be the same as all of the graduation marks. Visual angles are for longest anticipated viewing distance.

[2] 4.36 mrad (15 min) for light markers on dark background.

Figure 12.3 Graph design guideline from MIL-STD-1472G standard.

- The NHTSA (National Highway Traffic and Safety Administration) provided guidelines (Campbell 2018) in terms of icon and text size to display critical and non-critical information in a car display as follows:
  - Icon size
    - Optimal visual angle of primary graphical elements: 86 arcminutes (please refer to Chapter 2 for visual angle calculation)
    - Minimum visual angle of primary graphical elements
      - 41 arcminutes for time-critical applications
      - 34 arcminutes for non-time-critical applications
  - Text size (both within the icon and free-standing text)
    - Optimal height: 20 arcminutes
    - Minimum height: 16 arcminutes for time-critical applications; 12 arcminutes for non-time-critical applications
- The MIL-STD-1472G standard (2012) from the US Department of Defense provides detailed guidelines on electronic interface design, including font size, font type and icon size, and in different lighting conditions, graph display, and visual and auditory alert design, part of which is also inducted into NASA spacecraft cockpit design guidelines (Figure 12.3)

12.6 CASE STUDY: THE COMMON USER PROFILE

The last section of this chapter will present a case study on an ongoing effort to draft an ITU recommendation on common user profile for personalizing the presentation of audiovisual media (Biswas 2025). The rapid advance of technology and computing power has enabled users to access information through multiple devices, each having different sets of applications and software platform options. Developing different versions of the same content targeted at people with different needs in different situations is not scalable. Ways to adapt the same content for different users based on a user profile have already been explored in various research and commercial products, although a universal standard still does not exist considering the varieties of applications, devices and users' needs. A common user profile (CUP) is intended to personalize audiovisual content for a wide range of applications, devices and users, including users with disabilities. The CUP is targeted at broadband, digital TV, computer and smartphone software and web-based audiovisual systems. The CUP is independent of the device or application being used and is intended, through a software agent stored in the user's device or devices, to personalize the media experience by adapting the device's interface parameters such as the displayed captions' font size, colour and contrast; the audio dialogue balance, volume and equalization; and the size and position of onscreen elements. The CUP application can also be used to create synthetic profiles that can be used to simulate the effect of parameter adjustment on different devices, allowing developers to assess features on different persons with different ranges of abilities and contexts of use.

12.6.1 Objectives and structure of CUP

A user profile can be defined as an instance of a user model, and a user model can be defined as a machine-readable description of a user's needs. When applied to personalized accessibility, a common user profile can be said to have the following objectives:

1. Personalizing a user's device interface content and layout for different applications
2. Offering personalized accessibility services to all devices and platforms
3. Improving usability by sharing personalized content settings and interfaces across different platforms and devices
4. Adapting the quality of accessibility services (e.g., font size of captions) across multiple media applications
5. Secure sharing of personalization data between service providers or content developers
6. Enabling AI agents to automatically personalize or evaluate user interfaces

The CUP will be defined as a minimal list of variables that can be used to personalize user interfaces considering a wide range of devices and applications. While implementation will define the variables themselves, this standardization effort emphasizes the structure and requirements of describing the CUP. Structurally, three hierarchical levels are used to describe variables (Figure 12.4). The innermost layer is the baseline text, audio and video. The next layer describes the device media rendering options such as 2D or 3D display, speakers and haptic devices. It also integrates the device properties for the range of adjustable settings available to the profile. Finally, the outermost layer describes various accessibility services and their associated properties. Examples of a sample CUP code describing captions for 2D and 3D rendering are given in Figure 12.5.

A service provider can add additional variables required by a specific application or for future technology developments. Any user profile creation application should run locally to populate fields. It is not necessary to present all fields to the user separately. For example:

- Automated completion for colour blindness may predict a colour combination contrast
- Visual acuity can predict minimum font size and modality
- Audiometry data can anticipate relative dialogue to background balance and loudness

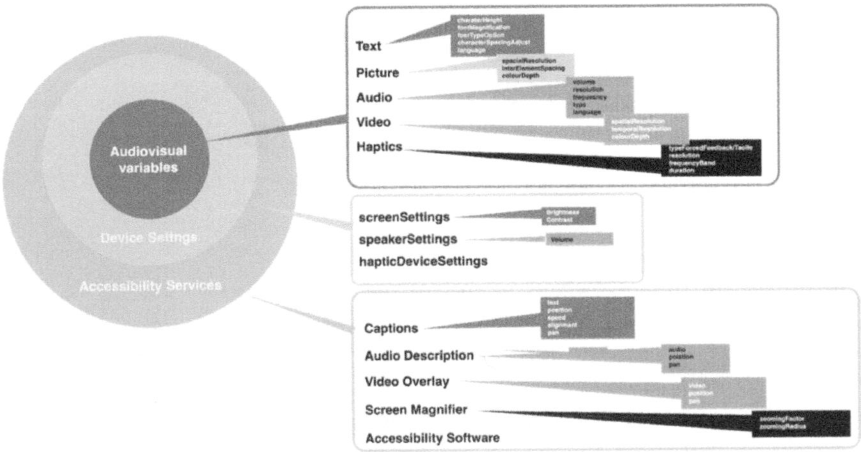

Figure 12.4 Hierarchy for describing variables.

Figure 12.5 Example profile for captions.

12.6.2 Implementation of CUP

In practice a user would set up a CUP by:

- Creating a profile using a profile creation application (PCA)
- Deciding to enable the option to upload the profile to a third-party cloud application. This is not mandatory and indicated by dashed lines in Figure 12.6
- Transfer the profile from the PCA and store it in a personal device (or in another location on the PCA device if both applications are on the same device)
- Create multiple profiles for different devices if needed

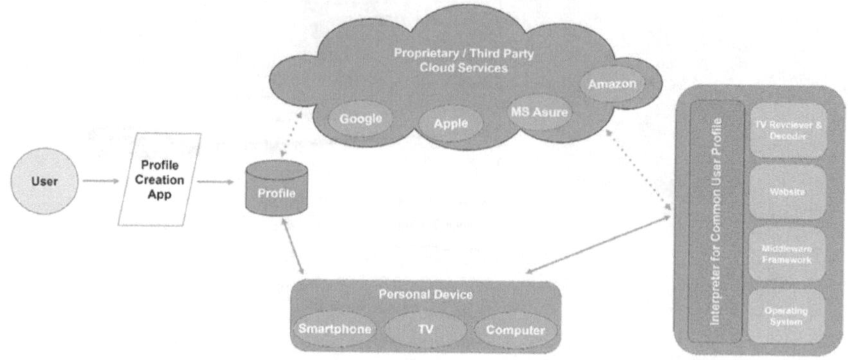

Figure 12.6 Sequence of CUP implementation actions.

After the user has completed the process:

- An interpretation application on the user's device will adapt the user interface based on the user's profile
- The interpretation application should understand the variables defined in the CUP in order to correctly interpret the personalization parameters for the content

12.6.3 Requirements for CUP implementation

Conformance to this recommendation requires common user profile specifying bodies to define a set of variables in both machine- and human-readable formats. Any variable defined shall include a permitted range of values, including their units of measurement.

- **Machine-readable format**: The recommendation requires common user profile specifying bodies to define the list of variables as part of metadata in XML or any other standardized format like JSON or other markup language format
- **Human-readable format**: The recommendation requires common user profile specifying bodies to define the list of variables and required value sets so that other stakeholders can understand and apply the requirements of the user profile

This recommendation only mandates that a description of any set of variables is given. It does not mandate or propose the exact format, naming conventions or specific units used to describe those variables. It also does not require enabling the end user to modify all variables – only a subset of variables can be exposed to end users to modify within a predefined limit set up by the device and application.

12.6.4 Mandatory requirements

A CUP specification should describe:

MR1. A JSON or XML schema that defines the set of variables of the CUP with permitted values and syntax

MR2. A set of human-readable descriptions of the variables defined in the schema

MR3. Variable descriptors for the following properties of textual information:
- Font property as described in https://www.w3.org/TR/css-fonts-3/#font-size-prop
- Character opacity as a percentage from 0% to 100%
- Foreground colour
- Background colour
- Standard or unit used to describe colour, which may be a single colour or gradient or shade of multiple colours based on rendering ability of a device
- Background opacity as a percentage from 0% to 100%
- The language of text specifies the xs:language attribute which will be as defined in W3C XML Schema Definition Language (XSD) 1.1 Part 2: Datatypes

MR4. A user profile description having a variable for each individual accessibility service. Accessibility services include but are not limited to caption, subtitle, audio caption and sign language

MR5. If a variable describes a caption or subtitle, the variable descriptor must have the following properties:
- Text of subtitle/caption specifies all variables defined in MSR5 to describe the text elements of subtitle/caption. Timing control of caption or subtitle. Position of caption or subtitle in 2D or 3D display. Variable describing textual information as mentioned in MR5. Position of caption or subtitle according to writing system
- Timing control of caption or subtitle
- Position of caption or subtitle in 2D or 3D display
- Variable describing textual information as mentioned in MR5
- Position of caption or subtitle according to writing system

MR6. All variables could be changed within a range of values following limitations imposed by a device, application and format. All changes of variables will maintain any semantic information conveyed by default settings of variable properties – for example, any critical information or warning will remain at a bigger font and in a different font colour than other information on display or the font size of subtitles will be changed within a limit that does not obscure main content on display. For IPTV applications, the recommendation requires supporting the profile changes described in Table 1 of H702

MR7. A user profile would depend on the media content. The user can define different user profile descriptions linked to the content. Alternatively, AI would understand the media content and alter the variables

12.6.5 Optional requirements

This section outlines options that are preferred but are not mandated:

OR1. The CUP may represent user choices for selecting between multiple caption tracks, including, for example, language, presence of hard of hearing text and simplified language

OR2. To develop a user profile creation application that collects users' preferences

OR3. A user profile description can optionally describe variables for:

- Icons
- Audio
- Video
- Haptics

OR4. Variables describing icons can optionally have following properties:

- Icon size
- Unit of icon size
- Background opacity as a percentage from 0% to 100%

OR5. Variables describing audio can optionally have the following properties:

- Loudness of audio signal
- Unit of loudness
- Resolution of audio
- Unit of resolution
- Audio coding format
- Language of audio
- Guiding mechanism for XR content such as an arrow to indicate the sound source either for lip reading or to locate the origin of the dialogue for the subtitles/captions

OR6. Variables describing video can optionally have following properties:

- Audio specifies all variables defined in OSR4 to describe the audio stream of the video
- Presence of caption/subtitle will be true or false based on the presence of a caption/subtitle in the video

- Complexity of video content will have a set of predefined values based on the speed of the content. For example, the complexity of a video depicting ice hockey can be marked as high visual load, while that of a news broadcast can be marked as low visual load
- Spatial resolution of video
- Unit of spatial resolution
- Video coding format
- Codec standard

OR7. Variables describing Haptics can optionally have following properties:

- Frequency of haptic feedback
- Duration of exposure to haptic feedback
- Location of haptic feedback
- Location of Haptic Feedback

Privacy considerations: A Common User Profile Document is categorized as sensitive PII for the purposes of establishing privacy safeguarding requirements and managing privacy risks.

12.6.6 Examples of CUP implementations

This section will demonstrate the use of CUP for a desktop and virtual reality (VR) application. The desktop application was a web-based and Android application for geotagging. The application offers tagging of exact location on Google Maps (Figure 12.7) overlaid with customized labels. The tagged data can be shared with other users who can access the exact longitude and latitude of the tags. The application also includes the current weather and forecasts for the next three days, and has current traffic layer options when zooming in to a given point or tag. For example, a travel website can allow different users to tag various locations such as ATMs, places to visit and medical centres that have accessible features or facilities. Similar applications are designed for reporting civic and healthcare issues for quick identification of serious problems. The geospatial and temporal clustering of information provided by elderly users will be useful for quick identification and resolution of critical civic and healthcare problems. The application is deployed as a webpage (https://cambum.net/CambUM/GeoTag/) and an Android app for smartphones (Figure 12.8).

The geotagging application can be used as a disaster reporting tool for users as well as the local, regional and national authorities. Users in affected areas or who can see the possibility of an incident, e.g., rising water levels, potential landslides and avalanches, can tag exact locations. This information can be accessed by local authorities and other users, which will help them maintain their safety or allow pre-emptive action to mitigate the impact. For a visually impaired user, the rule-based system selects a

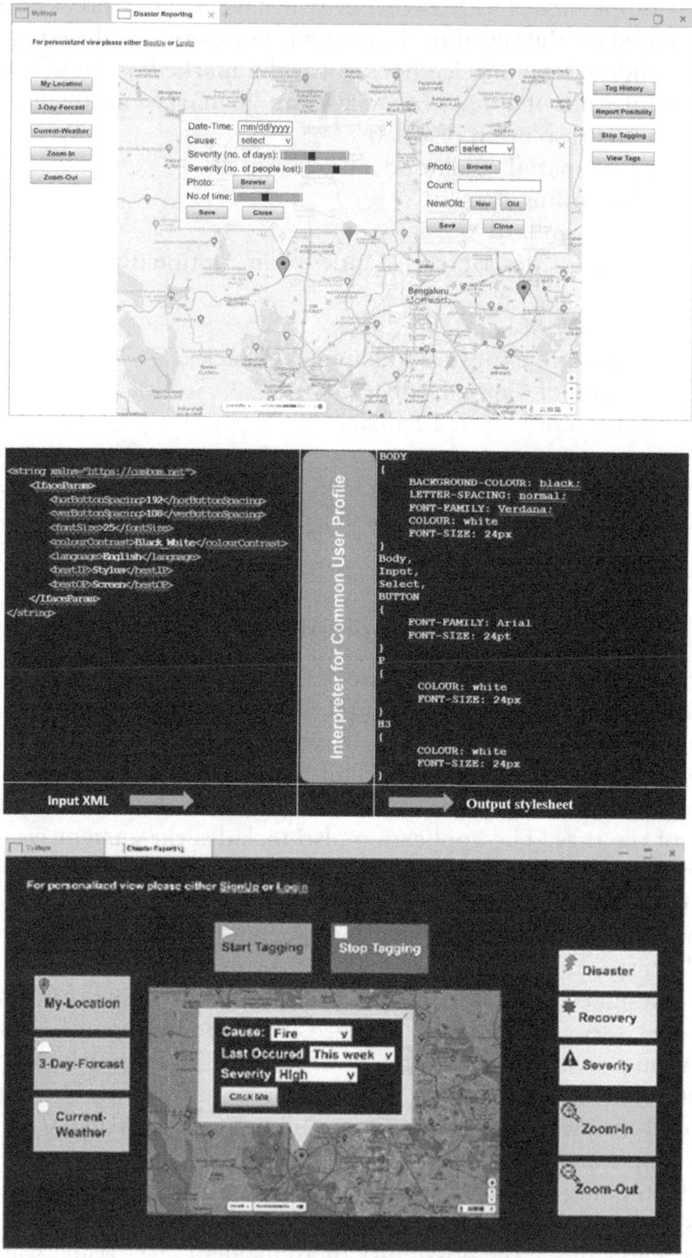

Figure 12.7 CUP example rendering implementation input and output with example render.

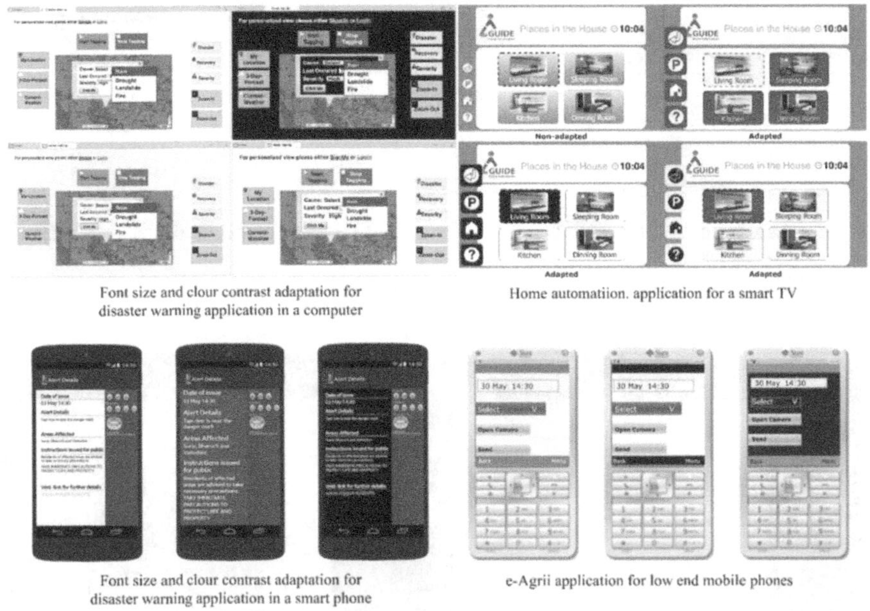

Figure 12.8 Example of interface adaptation for different devices.

stylesheet based on font size requirements and the type of colour blindness of users. The user profile creation application calculates and stores minimum visual angle requirements by estimating the distance to the screen from the type of device and screen resolution. For this application, the pop-up tagging windows are shown in yellow–blue colour contrast for red–green colour-blind users and in reverse contrast for any other type of colour blindness (Figure 12.8). This application can be further enhanced at any time by adding additional data to the map interface, which will allow many more user options to be included. Figure 12.8 shows examples of interface adaptation across multiple devices and platforms using a CUP format. The image colour, contrast, font size and inter-element spacing of icons can be adjusted across smart TVs, desktop and laptop computers, smartphones and low-end mobile phones based on an applied CUP.

A virtual reality application based on the SpaceX Dragon space shuttle modelled in Unity was used. The interior design (Figure 12.9) of the Dragon is well suited for this type of application as it is described as spacious, sleek and comfortable with touchscreens and a minimalistic control surface, which are highly intuitive and responsive, allowing astronauts to quickly access critical information and adjust as needed. The seating configuration can be customized for each mission. A CUP accessibility setting enables users of the VR application or virtual 'astronauts' with visual impairments to customize the displays to meet their specific needs, such as adjusting the

Figure 12.9 VR simulation SpaceX crew capsule.

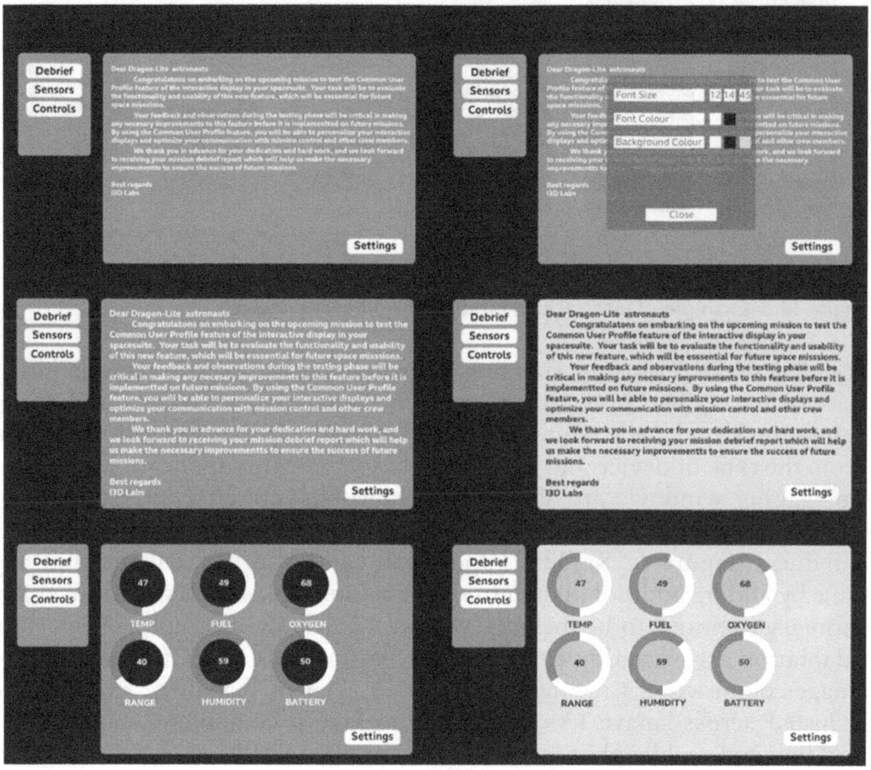

Figure 12.10 Simulation of CUP adaptation on the actual SpaceX crew capsule screens.

font size, contrast and colour scheme of the text on the screens. This helps ensure all crew members can read the displays and access critical information without difficulty. A CUP accessibility option would enable customization of the displays (Figure 12.10) to personal preferences such as the language, units of measurement, layout and accessibility. These adaptations would ensure all virtual crew members can easily navigate the displays and operate systems with confidence, regardless of their individual preferences, minimize the potential for errors and help astronauts carry out their mission with confidence and ease.

12.7 QUICK FACTS

- An **international standard** is a technical standard developed by one or more international standards organizations. International standards are available for consideration and use worldwide
- **ISO**: The International Organization for Standardization is an independent, non-governmental, international standard development organization composed of representatives from the national standards organizations of member countries
- **ITU**: The International Telecommunication Union is a specialized agency of the United Nations (UN) responsible for many matters related to information and communication technologies. It was established on 17 May 1865 as the International Telegraph Union, significantly predating the UN and making it the oldest UN agency
- **Metaverse**: An integrative ecosystem of virtual worlds offering immersive experiences to users that modify pre-existing and create new value from economic, environmental, social and cultural perspectives. A metaverse can be virtual, augmented, representative of or associated with the physical world
- **Common user profile**: An ongoing effort to define a minimal set of variables to personalize user interfaces irrespective of devices and applications for users with a wide range of abilities and for a wide range of usage scenarios

12.8 SUMMARY

In summary, this chapter aimed to give a brief overview of existing standards on ergonomics (ISO 9241), accessibility (ITU FGAVA, IRGAVA), user profiling (ISO FDAS, ITU CUP), Metaverse (ITU FG Metaverse) and detailed design guidelines from automotive (NHTSA) and aerospace (MIL-STD-1472G and NASA) standards. Readers should be aware of these guidelines while designing, developing, personalizing and evaluating user interfaces or interaction technologies. The chapter also aimed to outline examples of the standard development process from pre-standardization work in terms of collecting use cases, skeletonizing exact requirements for standardization and building consensus among member nations through mirror organizations. While in the networking sector, like 5G/6G-related communications, standards often lead to new research, in the user interface domain, adherence to accessibility or usability standards can lead to research on improving or inventing new modalities of interaction and personalization or accessibility services. In the upcoming summary, we put further pointers to existing work on standardizing novel interfaces like extended reality (XR) or haptics and new technology like cognitive load or distraction detection.

Chapter 13

Concluding remarks

So now we have reached the final chapter. Readers have gotten here in one of two ways: either they jumped to this place or they finished reading earlier chapters. We will start this chapter with a short summary.

There is an interesting phrase often attributed to a great physicist: 'Experts do not know everything, but they know where to look'. In today's age of ChatGPT and numerous online resources, we can rephrase it as 'Experts do not know everything, but they know the appropriate keywords for Google search or the best prompt for ChatGPT'. The statement can be interpreted as experts being able to identify the main challenge in designing, developing or realizing a product and service. They may not be able to solve it themselves or with the existing team, but they can identify the correct resource to solve the problem and engage accordingly. A solution is often efficiently reached when the problem is correctly formulated.

The field of intelligent interaction design covers a wide array of topics starting from human factors to AI/ML, UI/UX design, statistical hypothesis testing and standardization, setting aside the domain knowledge required to solve individual case studies. Each of these topics warrants a full course or dedicated book in its own right, and this book by no means desires to replace existing textbooks on AI/ML or UI/UX design. The book aims to introduce readers to the breadth of the field so that they can map their project or research on intelligent interaction design to the appropriate vertical. The book envisions that a reader will have a brief overview of all relevant fields and will be able to identify the right tools and algorithms for implementing intelligent user interface.

After introducing the field of intelligent user interaction, the book started with a discussion on human factors – we discussed theories of visual search, visual pathways, colour perception and auditory perception. The third chapter introduced artificial intelligence and machine learning, mainly highlighting different ways of knowledge representation and corresponding algorithms. The next chapter introduced novel and alternative modalities of interaction, highlighting the use of AI algorithms in interaction design. The next two chapters went into details of computer vision and information retrieval, highlighting how we analyze image or video data and textual

DOI: 10.1201/9781003646716-13

information. We also introduced the basic concepts of large language models and vision language models. Chapter 7 explained information visualization and the architecture of a visualization pipeline. We also discussed a user study to compare two-dimensional and three-dimensional visual primitives. The eighth chapter presented techniques of experiment design and statistical hypothesis testing. The chapter aimed to explain data preprocessing, appropriate statistical test selection and reporting results. The next two chapters described applications of intelligent user interfaces, including human–robot interaction and human–machine interfaces for automotive and aerospace platforms. Chapter 12 discussed related standards for user interface and interaction design. The following appendix proposed a tentative layout and list of equipment for setting up a lab on intelligent user interaction.

Appendix I

Setting up an IUI lab

How does one set up a lab for undertaking research on intelligent user interface (IUI)? As we saw in the chapters, the range of IUI is quite broad, from AR/VR systems and cyber-physical systems to core AI and computer vision topics. In this appendix, I share a tentative floor layout and list of equipment with processes in India in late 2025.

Figure A.1 shows a layout of the proposed lab. It consists of the following units:

1. **Visualization theatre**: For research on virtual reality systems, telerobotics workstation and large-scale data visualization like smart city, climate, molecular visualization and so on. The lab will consist of a large-sized video wall and surround audio system.
2. **Simulation lab**: It may consist of a physical simulator for cars, aircraft or spacecraft with a motion platform. Space is marked for storing GPU workstations to run visualizations for the physical simulator. It will be near the visualization theatre for post hoc analysis of the simulation scenario in large-scale display and setting up VR-based simulators as described in Chapter 10.
3. **Usability chamber**: This will be an interrogation chamber-type room with one-sided glass so that engineers or designers can observe users without disturbing human–machine interaction, and the chamber will be equipped with physiological sensors. The chamber will be soundproof, and the internal illumination should be controllable within a range of 10 lux to 10,000 lux with appropriate diffusers to simulate dark and sunlight conditions.
4. **Drone testing area**: The drone testing area will be a netted area for flying and testing nanodrones and research on human–drone interaction.
5. **Cobot testing area**: This will be a caged area for placing cobots or any fixed-base manipulator and undertaking research on human–robot interaction.

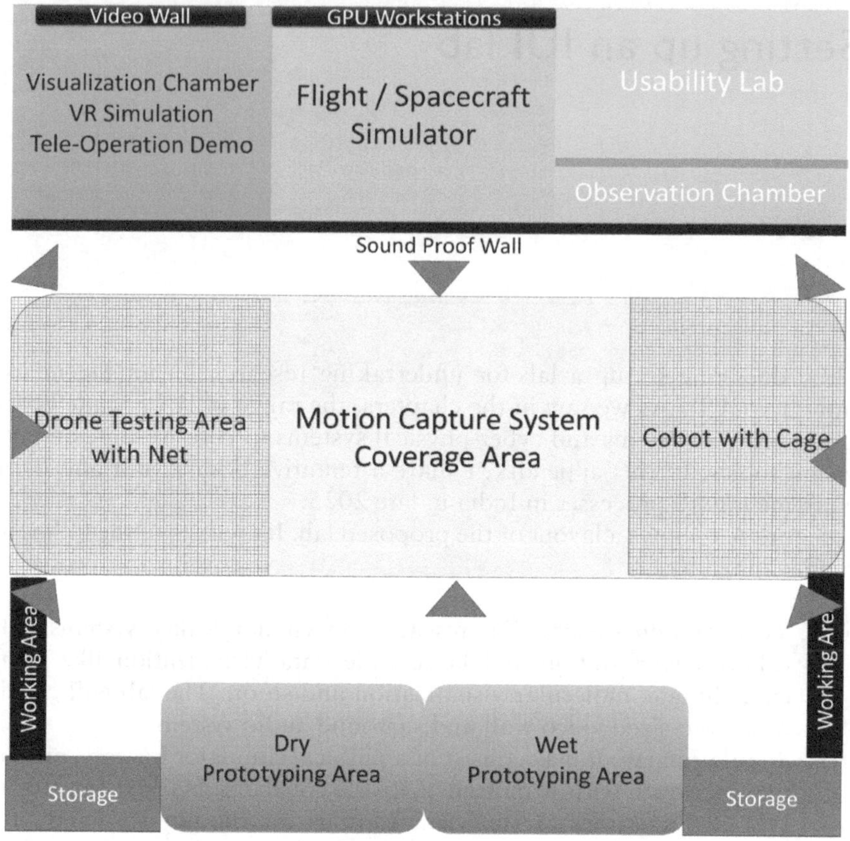

Figure A.1 Tentative lab layout for a lab on intelligent user interface.

6. **Motion capture area**: The central part of the lab will be empty and will be covered with motion capture cameras. It can be used to undertake research on motion planning for rovers, gait analysis of humans, validating object tracking for mixed or augmented reality systems and so on.
7. **Dry prototyping area**: It can house 3D printers and one or more spaces for placing personal computers for feeding models into the printer and a shouldering area for making mechatronics systems like customized grippers.
8. **Wet prototyping area**: This will be the only area with a connection to water and a drainage facility for using moulds or free-range design using clay or plaster of paris.

Table A.1 shows a tentative budget for equipment cost in India during 2025. The equipment is estimated for a class size of 20 users divided into using various facilities and also having their personal computing devices with access to standard reprographics facilities. The budget can be trimmed by removing equipment that does not have immediate requirements (like

Table A.1 Equipment list for a tentative lab on intelligent user interface

	INR (Lakh)		
Equipment	*Unit cost*	*Number of units*	*Total*
Workstation computer	4.5	5	22.5
GPU	4.5	5	22.5
AR/VR displays			
AjnaLens Headset	1	5	5
Meta Quest3	0.6	5	3
HTC Vive	4	1	4
Microsoft Hololens	5	1	5
MagicLeap2	1	7	7
iPad	0.5	5	2.5
Large display	0.4	5	2
Cyber-physical systems			
Low DoF robot	1.5	5	7.5
High DoF robot	5	1	5
Mobile bot/TurtleBot	1.5	2	3
Manipulators	2	6	12
Drones with SDK	10	2	20
DIY bots	0.05	10	0.5
Usability evaluation			
Screen-based eye tracker (Tobii Pro X series)	1	20	20
Eye-tracking glasses	1	40	40
EEG devices	1	10	10
Motion capture system	1	15	15
Simulator			
Basic Logitech simulator	1	1	1
Motion platform	1	20	20
Prototyping facility			
3D printers	4	0.6	2.4
Electronic toolbox	5	1	5
Mechanical toolbox	5	1	5
Total budget	**240 (Rs 2.4 crore)**		

eye-tracking glasses) or scaling down the number of units of 3D printers or GPU workstations. Similarly, the budget can be increased by adding more units to accommodate more users. The budget does not include any building material cost for making partitions, doing electrical wiring, or setting up cages or nets for drone and robot testing areas.

Appendix 2

List of software

1. Pointing Task Analysis: A web-based system to analyze and visualize ISO 9241 pointing task, discussed in Chapter 2 in context of Fitts' law.
   - http://cambum.net/I3D_PointingTask/index.html
2. Cambridge Simulator: A Windows-based usability evaluation tool, discussed in Chapter 2 in context of user modelling.
   - https://cambum.net/CambridgeSimulator.zip
3. Diffusion model for synthetic data generation: A software for synthetic data generation and associated machine tools dataset, as discussed in the case study section of Chapter 5.
   - https://github.com/i3dlabiisc/I3D-Tools-Dataset
4. COVID 19 Automatic Phase Detection: A web-based system to automatically divide the duration of COVID-19 spread, discussed in Chapter 7 in the context of data visualization for blind users.
   - https://cambum.net/Covid19/index.html
5. Multimodal HRI: Voice-guided object detection and real-time pick-and-place manipulation using a robotic arm.
   - https://cambum.net/I3DLab/Robotics.htm#SampleCodes
6. XR-based HRI: A basic code for developing a front end in Unity and controlling a DIY robotic manipulator from the Unity front end, discussed in the section on XR-based HRI in Chapter 9.
   - https://cambum.net/I3DLab/Robotics.htm#SampleCodes
7. LLM-based HRI: Sample code to develop an LLM frontend and assign a waypoint or target location based on natural language instruction or image input, discussed in the section on LLM-based HRI in Chapter 9.
   - https://github.com/i3dlabiisc/LLM-HRI
8. TLX and SUS: A Windows-based software to collect and calculate NASA TLX and IBM SUS scores, discussed in Chapter 9 in the context of usability evaluation and in Chapter 10 on case studies on cockpit design.
   - https://cambum.net/TLX_SUS.zip

9. Cognitive Assessment: A web-based cognitive testing program that can be used to undertake basic cognitive tests and was also used as part of the Voyager Display study during the Axiom 4 mission to the International Space Station.
   - https://cambum.net/cogTest/index.html
10. Speech Perception Test: A web-based auditory testing program, an online program to test speech perception with respect to native human British speaker as well as TTS for English and German.
    - https://cambum.net/SpeechPerception/index.html

Appendix 3

Sample questions

1. Compare and contrast the following:
   a. Serial and parallel visual search
   b. GOMS and SOAR user modelling systems
   c. Qualitative and quantitative data analyses techniques
   d. Designing UI using MS Visual Studio and Python QT
   e. Heuristic usability evaluation and cognitive walkthrough
2. Explain the spreading activation model in the context of ACT-R cognitive architecture.
3. Explain how Fitts' law is used to compare input pointing devices.
4. Find the mean, median, first quartile, third quartile, inner fence and outer fence for the following set of numbers: 34, 22, 67, 12, 55, 87, 23, 145, 32, 90, 89, 5.
5. Investigate existing user interfaces of software applications, websites, smartphones or any other electronic product (e.g., microwave oven, camera). Identify three problems with existing user interfaces. Propose a new design to address the problem. If possible, relate the problem to a human factor (e.g.,. visual search, rapid aiming movement, learning, memory).
6. Why is depth-first search not complete?
7. What is (are) the condition(s) for an A* search to be optimally efficient?
8. Write an expression and identify variables for the naïve Bayes model.
9. Define precision and recall in the context of information retrieval.
10. What is the sphericity assumption in the context of ANOVA?
11. What is the difference between parametric and non-parametric tests in the context of statistical hypothesis testing?
12. Write one advantage and one disadvantage of a wearable eye gaze tracker over a screen-mounted eye gaze tracker?
13. Draw a set of blocks showing basic computation of a gesture recognition system. The input to the system will be live video feed and output will be a semantic gesture such as tap or swipe.
14. What is the input and output of a convolutional neural network (CNN) in the context of object detection?
15. What is the advantage of K-medoids clustering over K-means clustering?

MULTIPLE CHOICE QUESTIONS

1. NASA TLX scale is used for:
   A. Cognitive load estimation
   B. EEG measurement
   C. Estimating users' preference
   D. User interface personalization

   Answer: A

2. What is the advantage of vibro-tactile haptic systems over force feedback systems?
   A. Less bulky
   B. Consume less power
   C. Portable
   D. All of the above

   Answer: D

3. A GOMS model can be used to:
   A. Predict task completion time
   B. Model expert performance
   C. Divide goal of interaction into subgoals
   D. All of the above
   E. Both A and B

   Answer: D

4. What is a correct expression of index of difficulty in the context of Fitts' law?
   A. Log (1 + 2 × Distance/Width)
   B. Log (1 + 2 × Width/Distance)
   C. Distance/Width
   D. All of the above

   Answer: A

5. When users are involved in complex tasks, the demand on ________ can be significant.
   A. Short-term memory
   B. Shortcuts
   C. Objects that appear on the screen
   D. All of the above

   Answer: A

6. Which type of visual search is faster?
   A. Parallel search
   B. Serial search
   C. Both have the same speed

   Answer: A

7. The chunking mechanism in SOAR cognitive architecture is related to:
   A. Learning
   B. Predicting task completion time
   C. Both A and B
   D. None of the above

   Answer: A

8. Augmentative and alternative communication aid is commonly used for:
   A. Pilots
   B. Drivers
   C. Blind users
   D. Users with speech and motor impairments
   E. All of the above

   Answer: D

9. What is the range of pupil dilation when the eyes are open?
   A. 2 to 8 mm
   B. 12 to 18 mm
   C. 1 to 5 cm
   D. 0.1 to 1.5 mm

   Answer: A

10. If the distance from the eye to target is D and the target size is S, which is the correct expression of visual angle?
    A. 2 arctan(S/2D)
    B. 2 arctan(D/2S)
    C. 2 arcsin(S/2D)
    D. 2 arctan(SD)

    Answer: A

11. The Midas touch problem in an eye gaze-controlled interface refers to:
    A. Accuracy of eye tracking
    B. Selecting screen elements
    C. Frequency of eye tracker
    D. Organization of screen elements
    E. Pointing to screen elements

    Answer: B

12. Which statement is true about Fitts' law?
    A. Predicts target width from movement
    B. Predicts movement time from target width and distance
    C. Predicts length of target from movement time
    D. All of the above

    Answer: B

13. What is the disadvantage of the GOMS model?
    A. Cannot predict task completion time
    B. Requires detailed parameter tuning
    C. Only model that assumes error-free performance
    D. All of the above

    Answer: C

14. What is the visual angle of a 2 mm vertical line from a distance of 65 cm?
    A. 65 arctan(2)
    B. arctan(0.2/65)
    C. arcsin(2/65)
    D. arctan(65/0.2)

    Answer: B

15. Which body part has the highest tactile spatial resolution?
    A. Finger tip
    B. Palm
    C. Hair
    D. Chest

    Answer: A

16. Find the index of performance for the following Fitts' law study results in bits/sec. The data can also be found at https://tinyurl.com/2p8ht4tf.

Width	*Distance*	*Movement time (msec)*
45	80	2442
45	160	3265
45	240	2745
45	325	2998
55	80	2398
55	160	2750
55	240	2759
55	325	1899
65	80	1950
65	160	1603
65	240	2409
65	325	2885
75	80	2387
75	160	1830
75	240	1977
75	325	2027

A. 0.02
B. 0.87
C. 0.44
D. 0.50

Answer: B

17. Stereoscopic displays are different from non-stereoscopic ones in terms of:
A. Resolution
B. Refresh rate
C. Motion blur
D. Depth perception

Answer: D

18. What is the accommodation vergence problem in the context of AR?
A. Less spatial resolution
B. Less temporal resolution
C. Motion blur
D. Adjusting perceived depth of synthetic objects with real objects

Answer: D

19. What is a non-rigid object in Unity?
    A. Deformable
    B. Breakable
    C. Both A and B
    D. None

    Answer: C

20. HMD stands for:
    A. Head-mounted display
    B. Head-masked display
    C. Head-made display
    D. Head-mounted detection

    Answer: A

21. Which is an example of a 3D display?
    A. iPhone
    B. Hologram
    C. Vuzix AR headset
    D. COTS automotive HUD

    Answer: B

22. Select the appropriate response to the following sentences:
    i. Visualization is meant to enhance human potential, not to replace it.
    ii. Visualization helps to identify hidden and undiscovered data features.
    iii. When creating visualizations, perception is not as crucial to take into account.
        A. False, true and false
        B. True, true and false
        C. True, false and false
        D. True, true and true

    Answer: B

23. What is the difference between a visualization chart and visual analytics tool? (Select all that apply.)
    A. Both are the same.
    B. A visual analytics tool represents data using graphical elements, while a visualization chart is a framework designed to explore, analyze and interpret data interactively.

C. A visual analytics tool is a software or framework designed to explore, analyze and interpret data interactively, while a visualization chart is a static or interactive representation of data using a graphical element.
D. A scatterplot is an example of a visualization chart; Power BI is an example of a visual analytics tool.

Answer: C and D

24. Isocontours and direct volume rendering are techniques to visualize what kind of data?
    A. Tabular data
    B. Multidimensional data
    C. Geometric data
    D. Scalar data

    Answer: D

25. Which are the programmable stages in the graphics pipeline? (Select all that apply.)
    A. Vertex shader
    B. Rasterization
    C. Fragment shader
    D. None of the above

    Answer: A and C

26. Select the appropriate response to the following sentences.
    i. A network is a hierarchical structure where every node has one parent and zero or more children.
    ii. Trees are acyclic.
    iii. Files and folders are examples of tree dataset type.
    iv. A network can represent complex interconnected systems.
        A. True, true, true and true
        B. False, true, false and true
        C. False, true, true and false
        D. False, true, true and true

    Answer: D

27. Which of the following is a key benefit of incorporating data visualization into user interface (UI) design?
    A. It simplifies the codebase, making the application faster.
    B. It allows users to interactively explore and understand complex data.
    C. It ensures consistent design across all UI components.
    D. It eliminates the need for user training and onboarding.

    Answer: B

28. In the context of user interface design, which of the following scenarios best illustrates the use of data visualization to enhance user decision-making?
    A. A pie chart embedded in a dashboard showing the percentage share of product sales
    B. An auto-complete text box allowing users to search for a product by name
    C. A dropdown menu with a list of selectable regions for viewing corresponding sales data
    D. All of the above

    Answer: A

10. Infrared-based imaging provides better accuracy than appearance-based gaze estimation.
    A. True
    B. False
    C. True only for indoor environments
    D. Can't say for sure

    Answer: A

29. Appearance-based gaze estimation techniques can be used for:
    A. Only for laptop cameras
    B. Only for mobile phone front cameras
    C. Any RGB camera
    D. Any camera image (RGB/NIR) where the end-to-end learning technique is adopted

    Answer: D

30. What is the part of the eye that allows light to pass through to the retina?
    A. Lens
    B. Iris
    C. Pupil
    D. Cornea

    Answer: C

31. Which of the following aspects affect the accuracy of Infrared-based gaze estimation?
    A. Iris colour
    B. Accuracy of pupil centre estimation
    C. Accuracy of retinal centre estimation
    D. Diameter of cornea

    Answer: B

32. When collecting a dataset to train a deep learning model to estimate eye gaze, which of the following factors is deemed most important?
    A. Number of image-label pairs/size of the dataset
    B. Number of subjects in the dataset
    C. Head pose variations in the dataset
    D. Wide distribution of parameters in the dataset

    Answer: D

33. In the aviation case study, what was the main limitation observed for the COTS eye tracker?
    A. Poor detection rate
    B. Wearing the eye tracker was difficult for the pilot
    C. Accuracy of the eye tracker
    D. COTS eye tracker processing speed was slow

    Answer: A

34. In order to obtain faster estimates from a deep learning model, which of the following metrics has to be low?
    A. Number of parameters of the model
    B. Number of convolutional layers
    C. Number of fully connected layers
    D. Number of batch normalization and dropout layers

    Answer: A

35. In the context of convolutional neural networks, advantages of dilated convolution over normal convolution are:
    A. Larger receptive field (i.e., no loss of coverage)
    B. No loss of resolution of the output image (as we dilate instead of performing pooling)
    C. Better time and space complexity
    D. All of the above

    Answer: D

36. Which of the following options best explains the term 'forward kinematics' with respect to robotics?
    A. Creating a record of robot poses for forward motion
    B. Finding the end effector position and orientation given the joint angles
    C. Creating a map of the robot environment for motion planning
    D. Finding the joint angles given the end effector position and orientation

    Answer: B

37. What is the primary advantage of Gaussian splatting in 3D reconstruction?
    A. It eliminates the need for camera calibration.
    B. It allows for continuous representation of volumetric data.
    C. It is only suitable for small-scale environments.
    D. It uses a voxel grid for data representation.

    Answer: B

38. Which aspect of Gaussian splatting makes it suitable for dynamic scene reconstruction?
    A. The ability to update Gaussian parameters in real time.
    B. Its reliance on dense depth maps for every frame.
    C. Its computational complexity, which is higher than voxel-based methods.
    D. Its focus on global rather than local optimizations.

    Answer: A

39. How does Gaussian splatting approximate high-resolution details?
    A. By using a large number of uniformly sized Gaussians
    B. By dynamically adjusting the covariance matrices of Gaussians
    C. By relying on pre-computed voxel grid densities
    D. By ignoring small variations in scene geometry

    Answer: B

40. Which statement is true about cognitive walkthrough?
    A. Evaluates cognitive load
    B. Experts simulate user interaction from high- to low-frequency tasks
    C. Predicts visual perception
    D. Predicts task completion time

    Answer: B

41. What is p in the context of statistical hypothesis testing?
    A. Probability of type II error
    B. Probability of type I error
    C. Probability of RMS error
    D. Percent of variance explained

    Answer: B

42. Which statement(s) are true about heuristic evaluation?
    A. Easy to use
    B. Requires multiple evaluators
    C. Both A and B
    D. Neither A nor B

    Answer: C

43. Which of the following fact(s) indicate(s) overfitting in a machine learning model?
    A. High error during training, low error during testing
    B. High error during training, high error during testing
    C. Low error during training, high error during testing
    D. Low error during training, low error during testing

    Answer: C

44. A box and whisker plot shows:
    A. Min, max, median, 1st and 3rd quartiles
    B. Min, max, average and standard deviation
    C. Min, max, average, 1st and 3rd quartiles
    D. All of the above

    Answer: A

45. Which test can be used as the non-parametric equivalent of one-way ANOVA?
    A. Signed-rank test
    B. t-test
    C. ANCOVA
    D. H test

    Answer: D

46. What is the sphericity assumption in ANOVA?
    A. Nearly equal variance of all columns
    B. Nearly equal mean of all columns
    C. Equal number of rows of all columns
    D. None of the above

    Answer: A

47. What is the correct expression of the result for the following ANOVA table:

Source of variation	*SS*	*df*	*MS*	*F*	*P-value*	*F crit*
Between groups	56784662	2	28392331	29.18051	1.16E–11	3.047605
Within groups	1.7E+08	175	972989.5			
Total	2.27E+08	177				

    A. $F(2,175) = 29.18$, $p < 0.01$
    B. $F(2,177) = 29.18$, $p < 0.01$
    C. $F(175,177) = 29.18$, $p < 0.01$
    D. $F(175) = 2839$, $p < 0.05$

    Answer: A

48. With reference to the ANOVA table in the previous question, how many independent conditions were there?
    A. 2
    B. 3
    C. 4
    D. 1

    Answer: B

49. Calculate upper outer fence for the following list of numbers: 10, 34, 53, 556, 453, 232, 5, 66.
    A. 45
    B. 32
    C. 550
    D. 231
    E. 1065

    Answer: E

50. Find mean and median of the following numbers: 23, 34, 26, 12, 56, 23, 34, 29.
    A. 29.625, 27.5
    B. 27.45, 30.543
    C. 25.55, 45.01
    D. None of the above

    Answer: A

Appendix 4

Basic Maths Requirement for Teaching

This appendix is written for instructors or faculty colleagues who will use this book for teaching. Although this book aims to give a high-level overview of recent AI/ML concepts to designers, but it is expected that the instructors are aware of basic mathematical concepts behind basic algorithms and models like Gradient Descent, CNN and t-test, ANOVA. The following list points out a few basic mathematical concepts, which readers can study from any online or offline resources. The lecture slides at https://cambum.net/PB/lecture-notes.php can also be used.

- Partial Differentiation and finding Maxima and Minima of function with multiple variables, will be good to be aware of Hessian Matrix for teaching Gradient Descent and Backpropagation in Chapter 3
- Distance Metrics - Minkowski Distance, Manhattan and Euclidean Distance Metrics for teaching K-means Clustering in Chapter 3
- Conditional Probability and basics of Bayesian Inferencing for teaching Sensor Fusion in Chapter 4
- Basics of Matrix Algebra and Convolution operation for teaching CNN in Chapter 5
- Vector Algebra and Linear Projection/Transformation for teaching Chapter 6
- Concepts of Normal Distribution, Central Tendency (Mean, Mode, Median), Variance and Standard Deviation, Concept of Correlation and Covariance for teaching statistical hypothesis testing in Chapter 8

Bibliography

Anderson, J. R., & Lebiere, C. (1998). *The Atomic Components of Thought.* Hillsdale, NJ: Erlbaum.

Arjun, S., Reddy, G. S. R., Mukhopadhyay, A., Vinod, S., & Biswas P. (2021). Evaluating visual variables in a virtual reality environment. 34th British Human Computer Interaction Conference Interaction Conference, BCS HCI 2021, pp. 133–138.

Babu, M.D., JeevithaShree, D.V., Prabhakar, G., Saluja, K.P.S., Pashilkar, A., & Biswas, P. (2019). Estimating pilots' cognitive load from ocular parameters through simulation and in-flight studies. *Journal of Eye Movement Research*, 12(3), 1–16.

Bahdanau, D., Cho, K., & Bengio, Y. (2014). Neural machine translation by jointly learning to align and translate. arXiv:1409.0473.

Belongie, S., Malik, J., & Puzicha, J. (2002). Shape matching & object recognition using shape contexts. *IEEE Transactions on Pattern Analysis & Machine Intelligence*, 24(24), 509–521.

Biswas, P. (2025). Requirements for the semantics of a common user profile used to personalize audiovisual media. International Conference on Communication Systems and Networks, COMSNETS, pp. 1250–1253.

Biswas P and Robinson P (2010). Evaluating the design of inclusive interfaces by simulation, Proceedings of the ACM International Conference on Intelligent User Interfaces (IUI) 2010.

Biswas, P., & Langdon P. (2012). Developing multimodal adaptation algorithm for mobility impaired users by evaluating their hand strength. *International Journal of Human-Computer Interaction*, 28(9), 576–596.

Biswas, P., & Langdon, P. (2013). Developing and validating a hearing impairment simulator. *Journal of Assistive Technologies*, 7(3), 160–171.

Biswas, P., & Langdon, P. (2014). Multimodal target prediction model. Conference on Human Factors in Computing Systems – Proceedings, pp. 1543–1548.

Biswas, P., Robinson, P., & Langdon, P. (2012). Designing inclusive interfaces through user modelling and simulation. *International Journal of Human-Computer Interaction*, 28(1), 1–33.

Biswas, P., & Robinson, P. (2009). Modelling perception using image processing algorithms. Proceedings of the 23rd British HCI Group Annual Conference on People and Computers: Celebrating People and Technology, pp. 494–503.

Biswas, P., Arjun, S., Murthy, L.R.D., Prabhakar, G., Sharma, V.K., & JeevithaShree, D.V. (2020). COVID 19 data visualization through automatic phase detection. *ACM Digital Government: Research and Practice (DGOV)*, 1(4), Article 25.

Burdea, G.C., & Coiffet, P. (2016). *Virtual Reality Technology*. Hoboken, NJ: Wiley.

Card, S., Moran, T., & Newell, A. (1983). *The Psychology of Human-Computer Interaction*. Hillsdale, NJ: Lawrence Erlbaum Associates.

Campbell JL, Brown JL, Graving JS, Richard CM, Lichty MG, Bacon LP, Sanquist T (2018). Human factors design guidance for level 2 and level 3 automated driving concepts (Report No. DOTHS 812 555). National Highway Traffic Safety Administration, Washington, DC.

Caron, M., Touvron, H., Misra, I., Jegou, H., Mairal, J., Bojanowski, P., & Joulin, A. (2021). Emerging Properties in Self-Supervised Vision Transformers 2021 IEEE/CVF International Conference on Computer Vision (ICCV). IEEE, pp. 9630–9640.

Collaborative Robot Technical Specification ISO/TS 15066. (2016). Rockwell Automation: http://me.umn.edu/courses/me5286/robotlab/Resources/12-TR15066Overview-SafetyforCollaborativeApplications-RobertaNelsonShea.pdf

Corbin, J. and Strauss, A. (2014) Basics of Qualitative Research: Techniques and Procedures for Developing Grounded Theory. SAGE Publications, Inc., London, p 456.

Devlin, J., Chang, M.-W., Lee, K., & Toutanova, K. (2018). BERT: Pre-training of deep bidirectional transformers for language understanding. ArXiv:1810.04805.

Dieter, S., & Höllerer, T. (2016). *Augmented Reality: Principles & Practice*. Delhi: Pearson Education India.

Dosovitskiy, A., Beyer, L., Kolesnikov, A., Weissenborn, D., Zhai, X., Unterthiner, T., et al. (2021). An image is worth 16 x 16 words: Transformers for image recognition at scale. International Conference on Learning Representations.

Duchowski, A.T. (2007). *Eye Tracking Methodology*. London: Springer.

Elliott, D., Helsen, W.F., & Chua, R. (2001). A century later: Woodworth's (1899) two-component model of goal-directed aiming. *Psychological Bulletin*, 127(3), 342.

Field, A. (2009). *Discovering Statistics Using SPSS*. Thousand Oaks, CA: Sage.

Fitts, P.M. (1954). The information capacity of the human motor system in controlling the amplitude of movement. *Journal of Experimental Psychology*, 47, 381–391.

Gaikwad, P., Mukhopadhyay, A., Muraleedharan, A., Mitra, M., & Biswas, P. (2023). Developing a computer vision based system for autonomous taxiing of aircraft. *Aviation*, 27(4), 248–258.

Goodfellow, I., Pouget-Abadie, J., Mirza, M., Xu, B., Warde-Farley, D., Ozair, S., Courville, A., & Bengio, Y. (2014). Generative adversarial nets. *Advances in Neural Information Processing Systems*, 27, 2672–2680.

Hampson, P., & Moris, P. (1996). *Understanding Cognition*. Oxford, UK: Blackwell Publishers Ltd.

He, K., Chen, X., Xie, S., Li, Y., Dollár, P., & Girshick, R. (2022). Masked autoencoders are scalable vision learners. Proceedings of the IEEE/CVF Conference on Computer Vision and Pattern Recognition, pp. 16000–16009.

Hick, W.E. (1952). On the rate of gain of information. *Journal of Experimental Psychology*, 4, 11–26.

Himanshu, M.M., Sharma, V.K., Sulthan, J., Atulkar, A., Bhathad, D., & Biswas, P. (2025). Unlocking potential: Gaze-based interfaces in assistive robotics for SSMI users. IEEE International Conference on Robotics and Automation (ICRA 25).

Holden K, Greene M, Cross V, Sandor A, Thompson S, Feiveson A, Munson B (2023). Effects of Long-duration Microgravity and Gravitational Transition on Fine Motor Skills, Human Factors 65 (6), 1046–1058.

ISO 9241-303:2011. Ergonomics of human-system interaction – Part 303: Requirements for electronic visual displays.

ISO/IEC 24756:2009. Information technology – Framework for specifying a common access profile (CAP) of needs and capabilities of users, systems, and their environments.

ISO 10218-1:2025. Robotics — Safety requirements Part 1: Industrial robots.

ISO/IEC 42001:2023. Information technology — Artificial intelligence — Management system.

John, B. E., & Kieras, D. (1996). The GOMS family of user interface analysis techniques: Comparison and contrast. *ACM Transactions on Computer Human Interaction*, 3, 320–351.

Kieras, D., & Meyer, D. E. (1990). An overview of the EPIC architecture for cognition & performance with application to human-computer interaction. *Human-Computer Interaction*, 14, 391–438.

Kieras, D. (1994). GOMS modeling of user interfaces using NGOMSL. Conference Companion on Human Factors in Computing Systems, pp. 371–372.

Krishnan, A., Himanshu, C.D, Atulkar, A.A., Bhatad, D., Kharsade, M.R., & Biswas, P. (2025). Human-in-the-loop spacecraft maneuvers: Investigating human factors through VR and machine learning. IEEE SPACE 2025.

Krishnan, A., Kharsade, M.R., & Biswas, P. (2024). Comparison of view orientation in manned spacecraft through virtual reality simulation. 2024 IEEE Space, Aerospace and Defence Conference (SPACE), Bangalore, India, pp. 820–823.

Lindsay, R.K., Buchanan, B.G., Feigenbaum, E.A., & Lederberg, J. (1993). DENDRAL: A case study of the first expert system for scientific hypothesis formation. *Artificial Intelligence*, 61(2), 209–261.

Luck, S.J., Chelazzi, L., Hillyard, S.A., & Desimone, R. (1997). Neural mechanisms of spatial selective attention in areas V1, V2, & V4 of macaque visual cortex. *Journal of Neurophysiology*, 77, 24–42.

Mayor, A. (2018). *Gods and Robots*. Princeton, NJ: Princeton University Press.

McCulloch, W.S., & Pitts, W. (1943). A logical calculus of the ideas immanent in nervous activity. *Bulletin of Mathematical Biophysics*, 5(4), 115–133.

Mikolov, T. (2012). Statistical language models based on neural networks. PhD Dissertation at BRNO University of Technology.

MIL-STD-1472G, 11 January 2012, US department of defense, design criteria standard, human engineering.

Mitra, M., Kumar, G., Chakrabarti, P.P., & Biswas, P. (2024). Enhanced human-robot collaboration with intent prediction using deep inverse reinforcement learning. Proceedings – IEEE International Conference on Robotics and Automation, pp. 7880–7887.

Mitra, M., Kumar, G., Chakrabarti, P.P., & Biswas, P. (2025). Investigating inverse reinforcement learning during rapid aiming movement in extended reality and human-robot interaction. *ACM Transactions on Human-Robot Interaction*. Advance online publication.

Mitra, M., Pati, P., Sharma, V.K., Raj, S., Chakrabarti, P.P., & Biswas, P. (2023). Comparison of target prediction in VR and MR using inverse reinforcement learning. International Conference on Intelligent User Interfaces, Proceedings IUI, pp. 55–58.

Moore, B.J. (1997). *An Introduction to the Psychology of Hearing*. New York, NY: Elsevier Science Publishing Co Inc.

Newell, A. (1973). You can't play 20 questions with nature and win. Projective comments on the papers of this symposium. Pittsburgh, PA: Carnegie Mellon University, Department of Computer Science.

Newell, A. (1990). *Unified Theories of Cognition*. Cambridge, MA: Harvard University Press.

Nixon, M., & Aguado, A. (2002). *Feature Extraction & Image Processing*. Oxford: Elsevier.

Norman, K. (2017). *Wiley Handbook of Human Computer Interaction*. Hoboken, NJ: Wiley.

Olsen, A. (2012). The Tobii I-VT Fixation Filter. https://www.tobiipro.com/site-assets/tobii-pro/learn-and-support/analyze/how-do-we-classify-eye-movements/tobii-pro-i-vt-fixation-filter.pdf, Accessed on 05/04/2019.

OpenAI. (2023). GPT-4 Technical Report. ArXiv:2303.08774.

Peters, M., Neumann, M., Iyyer, M., Gardner, M., Clark, C., Lee, K., & Zettlemoyer, L. (2018). Deep contextualized word representations. *Proceedings of the 2018 Conference of the North American Chapter of the Association for Computational Linguistics: Human Language Technologies*, 1, 2227–2237.

Prabhakar, G., Ramakrishnan, A., Madan, M., Murthy, L.R.D., Sharma, V.K., Deshmukh, S., & Biswas, P. (2020). Interactive gaze and finger controlled HUD for cars. *Journal on Multimodal User Interfaces*, 14(1), 101–121.

Radford, A., Narasimhan, K., Salimans, T., & Sutskever, I. (2018). Improving language understanding by generative pre-training. OpenAI. https://www.mikecaptain.com/resources/pdf/GPT-1.pdf

Raj, S., Beri, N., Patel, D.S., Sinha, Y., Chakrabarti, A., & Biswas, P. (2024). HaM3D: Generalized XR-based multimodal HRI framework with haptic feedback for industry 4.0. *Journal on Multimodal User Interfaces*, 18(4), 331–349.

Russell, S., & Norvig, P. (2009). *Artificial intelligence: A modern approach* (3rd. ed.). Upper Saddle River, NJ: Prentice Hall Press.

Sahu, Y., Mukhopadhyay, A., Kumar, G., Kumar, A., & Biswas, P. (2024). A comparative study on image translation GAN models to improve object detection accuracy on low-resource domains. International Conference on Vehicular Technology and Transportation Systems (ICVTTS).

Salton, G., Wong, A., Yang, C.S. (1975). A vector space model for automatic indexing. *Communications of the ACM*, 18(11), 613–620.

Sharma, V.K., Murthy, L.R.D., & Biswas, P. (2022). Comparing two safe distance maintenance algorithms for a gaze-controlled HRI involving users with SSMI. *ACM Transactions on Accessible Computing*, 15(3), Article 27.

Shneiderman, B. (2016). *Designing The User Interface – Strategies for Effective Human-Computer Interaction.* Delhi: Pearson Education.

Shortliffe, E.H., & Buchanan, B.G. (1975). A model of inexact reasoning in medicine. *Mathematical Biosciences*, 23(3–4), 351–379.

Singh, U., Jeevithashree, D.V., Muraleedharan, A., Saluja, K.P.S., & Biswas, P. (2024). Accessibility analysis of educational websites using WCAG 2.0. *ACM Digital Government: Research and Practice*, 5(3), 1–28.

Sinha, Y., Shanmugam, S., Sahu, Y.K., Mukhopadhyay, A., & Biswas, P. (2025). Diffuse your data blues: Augmenting low-resource datasets via user-assisted diffusion. International Conference on Intelligent User Interfaces, Proceedings IUI, pp. 538–552.

Sutskever, I., Vinyals, O., & Le, Q.V. (2014). Sequence to sequence learning with neural networks. Advances in Neural Information Processing Systems, pp. 3104–3112.

Talwar, N., Prasad, H., Pena-Rios, A., & Biswas, P. (2025). Walking trajectory prediction using signal processing & machine learning based models. 10th IEEE International Conference on Advanced Robotics and Mechatronics (ARM 25).

Tovee, M.J. (2008). *An Introduction to Visual Systems.* Cambridge: Cambridge University Press.

Truitt, E.R. (2015). *Mediaeval Robots.* Philadelphia, PA: University of Pennsylvania Press.

US Department of Transportation, NHTSA, Human Factors Design Guidance for Level 2 and Level 3 Automated Driving Concepts.

Vaswani, A., Shazeer, N., Parmar, N., Uszkoreit, J., Jones, L., Gomez, A.N., Kaiser, T., & Polosukhin, I. (2017). Attention is all you need. Advances in Neural Information Processing Systems. 30. Curran Associates, Inc.

Weizenbaum, J. (1966). ELIZA–A computer program for the study of natural language communication between man and machine. *Communications of the ACM*, 9, 36–45.

Zhang, A., Zachary, C.L., Li, M., & Smola, A.J. (2023). *Dive into Deep Learning.* Cambridge: Cambridge University Press.

Zong, J., Pineros, I.P., Chen, M., Hajas, D., & Satyanarayan, A. (2024). Umwelt: Accessible structured editing of multi-modal data representations. In Proceedings of the 2024 CHI Conference on Human Factors in Computing Systems (CHI '24). Association for Computing Machinery, New York, NY, Article 46, pp. 1–20.

Zong, J., Lee, C., Lundgard, A., Jang, J., Hajas, D., & Satyanarayan, A. (2022). Rich screen reader experiences for accessible data visualization. Computer Graphics Forum, 41, 15–27. https://doi.org/10.1111/cgf.14519

Index